SEISMIC JAPAN

SEISMIC JAPAN

The Long History and Continuing Legacy
of the Ansei Edo Earthquake

Gregory Smits

University of Hawai'i Press
Honolulu

Library of Congress Cataloging-in-Publication Data
Smits, Gregory, 1960– author.
Seismic Japan : the long history and continuing legacy
of the Ansei Edo earthquake / Gregory Smits.
pages cm
Includes bibliographical references and index.
ISBN 978-0-8248-3817-1 (cloth : alk. paper)
1. Earthquakes—Japan—History. 2. Earthquakes in literature.
3. Ansei Edo Earthquake, Japan, 1855. I. Title.
QE537.2.J3S54 2013
551.220952—dc23
2013010525

ISBN: 978-0-8248-9417-7 (paperback)

Cover art: In this print from 1855, the "Edo" catfish in the foreground
represents the 1855 Ansei Edo earthquake; the "Shinshū" (Shinano) catfish
in the background represents the 1847 Zenkōji earthquake. Reprinted
with permission of the Royal Ontario Museum © ROM.

Cover design by Mardee Melton
Interior design by Janette Thompson (Jansom)

CONTENTS

PREFACE

The primary purpose of this book is to contribute to the study of Japanese history. My intended audience is scholars of Japan, especially historians. The topic of earthquakes will also appeal to readers with little or no knowledge of Japan. It is my intention that readers unfamiliar with Japan but who possess an interest in earthquakes, the history of science, the social history of catastrophes, and related topics will benefit from this book and be able to comprehend it without excessive struggle. For these readers I have made a point of defining or translating all Japanese terms and preferring translated English titles of Japanese works in the main text.

Popular culture in early modern Japan relied heavily on plays on words. Such wordplay was possible because of the many homonyms in Japanese and because of flexibility in the writing system. Writers could deploy combinations of Chinese characters and native syllabic scripts (essentially an alphabet) to facilitate wordplay or to add additional levels of nuance or humor. When discussing such matters, extensive use of Japanese terms is inevitable. Nevertheless, even in such discussions, attentive readers who do not know Japanese should be able to grasp the main points.

Many of the sources for this study are visual in nature. I have tried to write about prints and other visual images as richly as possible to convey a sense of their contents through words. Some essential material is reproduced within the book as figures. To moderate production costs, however, many images discussed can be accessed by typing into a browser the URL (universal resource locator) found in the notes. In selecting these URLs, I have favored well-established, large digital archives whose Web addresses are unlikely to change. Moreover, all other factors being similar, I have favored the shortest URL. Even if a URL does become inoperative over time, it is a simple task to search for a print using its Japanese name.

ACKNOWLEDGMENTS

It is a pleasure to acknowledge friends, colleagues, and support staff who have sustained and facilitated the production of this book. The process leading to this book began in 2004, when I was fortunate to collaborate with Ruth Ludwin on an article. Starting out as a side project, the work I did for our coauthored piece caused me to set aside past lines of research and shift most of my attention to earthquakes.

My work on earthquakes benefited from friends and colleagues who have sent me articles form a variety of media sources over the years. These contributors include Laura Nenzi, Erica Brindley, Jessamyn Abel, and Jonathan Abel. My wife, Akiko, has been especially helpful in sending me earthquake-related articles and in many other ways. Brian Atwater and Robert Geller have provided valuable assistance with respect to questions about seismology. I am especially grateful to Robert Geller for his fast responses to my questions and for providing me with scientific journal articles on a variety of key topics that I would not likely have encountered on my own. Some of this material has informed the present book, and all of it will be useful for future projects. I am grateful to Noriko Itasaka for her assistance in obtaining primary sources connected with the Ansei Edo earthquake.

The Pennsylvania State University has been instrumental in facilitating my research and writing. A year of sabbatical leave in 2004–2005 provided the ideal incubation conditions, ultimately resulting in this book and several articles on related topics. I am grateful to Gregg Roeber, Eric Hayot, and Michael Kulikowski for providing administrative assistance that facilitated my work. I have imposed a heavy load on the unsung heroes of Penn State's interlibrary loan staff, who have always been helpful and efficient. I would also like to thank the staff of the Tokyo Municipal Library, Japan's National Diet Library, and the Saitama Prefectural Museum of History and Folklore for a variety of assistance. Digital text and image collections available to the public have contributed greatly to this book. In particular, I would like to acknowledge Waseda University Library Catalog, the Earthquake Research Institute, University of Tokyo, and the Kokuritsu Kokue Toshokan digitalized sources collection at the National Diet Library.

I appreciate the opportunity to present my research and thoughts at a variety of conferences and symposia in recent years, and my ideas have benefited from discussions with Gregory Clancey, Jordan Sand, Daniel Botsman, Gerald Figal, Phil Brown, Garrit Schenk, Noura Dirani, Monica Juneja, Steve Phipps, Frank Chance, Kristina Buhrman, Haruko Wakabayashi, Bettina Gramlich-Oka, Amanda Stinchecum, Fabian Drixler, Ian Miller, and others.

Two reviewers for the University of Hawai'i Press provided thoughtful comments, helpful references, and useful criticism. I appreciate the time and effort you put into your anonymous work and regret being unable fully to implement all of your suggestions. I thank University of Hawai'i Press executive editor Patricia Crosby for her support for this project and expert guidance. It was a pleasure working with production editor Stephanie Chun and copy editor Lee Motteler. My brother Jeff, proprietor of Cherokee Drafting Specialists, produced the two maps for this book.

Finally, I would like to thank Paul Roomsburg and those who periodically gather at his cabin to play music, as well as my larger network of music friends. You help me keep the peculiarities of academic life in reasonable perspective, which helps make books like this one possible.

CHAPTER 1

Earthquakes in the Early Modern Era

Early in 2010, reports of strange fish came to the attention of Japan's news media. Large numbers of slender oarfish (*Regalecus glesne* or *Regalecus russelii*), commonly known in Japan as "messengers from the undersea dragon palace," had been showing up in shallow-water nets or washing ashore along the Sea of Japan coast. This denizen of deep water might be a harbinger of a big earthquake, wrote the *Kyodo News*. Basing this specula-tion on "an old saying," the report quoted "a specialist in ecological seismol-ogy" to the effect that deep-sea fish are especially sensitive to the movements of faults.[1] In the wake of the Great East Japan Earthquake and tsunami of March 11, 2011, alleged connections between aquatic animals and earth-quakes appeared in a variety of sources. One newspaper article pointed out that unusually large squid catches preceded several major earthquakes between 1946 and 2011 and speculated that squid might somehow function as precursors of earthquakes.[2] A pamphlet produced by the Kashima Shrine in an attempt to put the best possible face on the 3/11 disaster claimed that deities sent warning of the impending earthquake a week in advance by causing a pod of whales to appear off the coast near the shrine.[3] More recently, a newspaper article speculated that the frequent appearance of deep-sea oarfish along the coast of Shizuoka Prefecture has prompted con-cern that this "messenger from the undersea dragon palace" might signal an impending earthquake.[4] Claims of unusually large fish catches have been a regular feature following major earthquakes. Residents along the Sanriku (northeast) coast, for example, reported this phenomenon prior to the great tsunamis of 1896 and 1933, which some writers regarded in retrospect as precursors to these seismic events.[5] Recently, the City of Susaki in Kōchi Prefecture (Shikoku) announced plans to monitor animal behavior and well water levels and broadcast earthquake and tsunami evacuation recommendations should any anomalies be detected.[6] These recurring fish and animal tales are part of a continuing legacy of the 1855 Ansei Edo earthquake.

The Ansei Edo earthquake not only cast a long shadow into the modern era, it was also the product of a long history of early modern earthquakes. At first glance, it may seem odd to speak of an earthquake as a product of history. Here I refer to human history, not to the well-entrenched but increasingly controversial notion in seismology of recurring "characteristic earthquakes."[7] As a social phenomenon, earthquakes in Japan were meaningful events that demanded interpretation. By the middle of the nineteenth century, a rich tradition of reading earthquakes was in place. This process of interpretation, its rhetoric, and its history reveal much about cultural values and tensions and prevailing or competing worldviews. Similarly, the details of reactions to specific earthquakes can serve as unique windows on society, providing insights into issues of the day and relations between social groups.

When a major earthquake shook Edo on the night of November 11, 1855, writers and commentators immediately began assessing its meaning and significance. In many respects, this process followed past examples. One manifestation was a focus on Japan's geographic and cultural contours. Seismic activity long before, for example, supposedly rent the earth one night to create Lake Biwa in Ōmi Province and Mt. Fuji in Suruga Province. Such was the claim of *Ways of Earthquakes in Japan* (*Honchō jishin no shidai,* 1855 or 1856). Not only did earthquakes create iconographic geography, the text explains, they have periodically shaken the Japanese islands since the time of the earliest human emperors.[8] Earthquakes also shaped Japan's literary culture. *Land and Sea Earthquake Record* (*Kainai jishinroku,* 1855 or 1856), for example, includes a set of earthquake-related verses attributed to the "thirty-six superlative poets," including such cultural luminaries as Manyōshū poet Kakinomoto Hitomaro, early Heian poet Ono no Komachi, and mid-Heian poet Izumi Shikibu. Many of the verses in the set indeed deal with earthquakes, although the connection in some of the poems is remote.[9] On the ground, the shaking in 1855 brought older, pre-Tokugawa geographic contours of Edo into relief. Most dramatic was a partial reemergence of Hibiya Cove, filled in during the first decade of Tokugawa rule to create real estate for daimyō residences and government buildings. Mansions and bakufu offices located on land that had once been shallow ocean collapsed and burned so dramatically relative to surrounding areas that several observers made the connection between the former cove and the current destruction. Major earthquakes possessed the power to reveal and revive past geography, both real and imagined.

Earthquakes also had the power to reveal alternative political geographies. Throughout the early modern era, the experience of major earthquakes reinforced the idea of Japan as a temporal and spatial entity. Most earthquakes highlighted a political geography based on the classic sixty-six provinces, a temporal order measured in imperial reigns, and a land that was home to myriad deities led by Amaterasu of Ise. Just as the 1855 earthquake highlighted an older version of Edo's geography, most early modern seismic upheavals temporarily highlighted an older Japan organized in terms of imperial provinces, imperial time, and imperial deities. Only when the city of Edo shook violently in 1703, and especially in 1855, did a state consisting of a bakufu (shogunate) and domains—what modern historians often call the *bakuhan* state—figure prominently in the post-earthquake discourse. Otherwise, the Japan that shook was usually a land consisting of an imperial court and provinces.

In addition to foregrounding visions of Japan in which the *bakuhan* state was largely irrelevant, the destructive and terrifying aspects of major early modern earthquakes stimulated attempts to explain and understand the mechanical processes behind these events. Owing to empirical observation, academic knowledge of earthquakes accumulated throughout the Tokugawa period. By the early nineteenth century, discussions of epicenters, aftershocks, uplift, subsidence, liquefaction, severity of ground motion as a function of soil base, and other basic geophysical concepts and phenomena could be found in both academic and semipopular books. Diverse earthquake-related folklore also accumulated along with these advances in scientific knowledge. From the mid-seventeenth century, for example, earthquakes became associated with flashes of light, either in the sky or emanating from the ground. Observers and writers also commonly linked earthquakes with certain types of atmospheric phenomena and weather conditions. In 1855 the idea that catfish could help predict earthquakes emerged in a collection of unverified popular tales. Since then, this idea has taken such strong root that between 1977 and 1993, 120 million yen of taxpayer money went to fund catfish in government aquariums as an "advance guard of earthquake prediction," always on duty.[10] Both early modern scientific knowledge and the broader accumulated body of earthquake lore helped shape the development of seismology during the Meiji period and in some respects to the present day.

The sudden destructive power of violent ground movement exacted a psychological toll on those who survived. Severe earthquakes and their aftershock sequences produced anxieties about the viability of society itself. One reaction to these anxieties was a rhetoric of reassurance in the written materials that inevitably appeared soon after major earthquakes. These texts sometimes sought to explain the mechanics of earthquakes, usually in terms of yang energy trapped within the earth seeking to escape upward. Moreover, they often pointed out Japan's long history of earthquakes, upheavals that occurred even during the reigns of sagacious rulers. The ultimate point was resiliency. Past earthquakes did not destroy society and neither would the current calamity. Furthermore, this literature typically portrayed earthquakes as worldwide phenomena. While the main point was to make earthquakes appear normal, this approach also had the effect of situating Japan within a world of other countries. In this context, China was the main point of comparison during the early modern era, and Italy began to serve a similar function during the Meiji period.

By the early modern era, one common component of popular notions of Japan's distinctive qualities was that Japan was blessed with divine favor. The proof was the thousands of *kami* and buddhas throughout the Japanese islands. Bountiful harvests were a classic manifestation of the benevolence and blessing of these supernatural entities. Within Japan, the effects of earthquakes often invoked discussion of deities and cosmic forces. In this way, earthquakes amplified a vision of Japan as a land distinguished by its host of buddhas, bodhisattvas, and deities, tapping into a long, complex history of *shinkoku* (country of deities) discourse.

In this book, I argue that the Ansei Edo earthquake played a pivotal role in a process of shaping conceptions of Japan in the realms of politics, religion, geography, and natural science. Moreover, this earthquake produced new ideas about human agency vis-à-vis earthquakes that have affected notions of seismicity and society in modern Japan. One reason the Ansei Edo earthquake was so influential was accidental factors of time, place, and circumstance. The shaking of the shogun's capital coincided with the heightened religious fervor of an *okage* (by the virtue of) year, a year of special religious significance in a twelve lunar year cycle. This earthquake occurred at the beginning of a period of major challenges for the bakufu, and it weakened shogunal power in subtle but significant ways. Furthermore, because its occurrence was theoretically impossible according to prevailing theories of earthquake mechanics, Ansei Edo opened doors to alternative theories of seismic upheaval in academic circles. Lore that emerged from this event, such as the idea that earthquakes disrupt magnetic fields, continues to inform quests by amateurs and scientists seeking ways to predict earthquakes by studying alleged precursors.[11]

The Global Context

Although this study concerns itself with Japan, insights from scholars of disasters working in other contexts contribute to the analysis. It is also useful at the start to plot major changes in attitudes concerning earthquakes in other parts of the world to contextualize the Japanese situation. The approximate trajectory of understanding earthquakes in Japan matches that of many parts of Europe and the Islamic world. Until the early seventeenth century in all of these places, earthquakes were primarily religious events in terms of their cause and significance—usually considered divine punishment. During the eighteenth century, the situation became more complex, and mechanical or scientific explanations emerged as alternatives to religious explanations. By the nineteenth century, mechanical theories of earthquakes were well established around the world, but they did not displace or eliminate moral and religious explanations. Indeed, interpreting natural disasters as divine retribution remains common to this day throughout the world.[12]

Examples of earthquakes explained in terms of religion and morality abound in the premodern world. In classical Chinese thought, earthquakes were part of a range of abnormalities such as floods, droughts, and epidemic disease that might indicate moral or ritual deficiency in kings or rulers. The same thinking applied to smaller units within the Chinese Empire. In an 1149 manual on agriculture, for example, Chen Fu explains that "the deities of mountains and rivers" cause or prevent natural disasters. Therefore, one key to prosperity in agricultural villages was to diligently perform rites to these deities in a correct manner.[13] As we will see, Chinese ideas about natural hazards and disasters informed early modern Japanese thinking about earthquakes, albeit with some important differences.

In the Islamic world, earthquakes might serve either as divine punishment or divine warning, depending on the context. One prominent theme was that earthquakes were the wages of sin and immorality. Writing in Cairo in 1576 in the wake of an earthquake, Ibn al-Jazzār emphasized this theme, specifying relevant misdeeds, starting with the recreational consumption of a drug called *bursh* and dwelling on coffeehouse culture. He characterized these establishments as dens of vice featuring music, singing, and men ostensibly seeking coffee but really present owing to an "appetite for the beardless cupbearer." He recommended prayer and fasting as the proper responses to the earthquake.[14] Here we see the same rhetorical technique that some Japanese commentators on earthquakes deployed in both early modern and modern times: using the occurrence of a recent earthquake to amplify a writer's social criticism or agenda.

Interestingly, early Islamic earthquake texts sometimes mention a giant fish as a possible cause of earthquakes. The Egyptian writer Abū 'l-Fadl al-Suyūtī (1445–1505) listed thirty causes of earthquakes, one of which was that "Satan makes the fish on which the earth stands feel proud, so it moves." Ibn al-Jazzār also mentioned fish as a possible earthquake cause, although his focus was on condemning popular culture.[15] Connections between giant fish and earthquakes can be found in the folklore of many parts of the world. Although in some cases this lore migrated from one place to another (China to Japan, for example), it is likely that the movements and other attributes of aquatic creatures, especially in gigantic, mythical form, independently struck many people around the world as a suitable symbol of the shaking associated with earthquakes or the power to cause such events.

Typical premodern interpretations of earthquakes in areas dominated by Christianity similarly regarded them either as divine punishments or, especially when the level of destruction was low, divine warnings. Following a French earthquake in 1549, for example, the Montélimar town council prohibited dancing in connection with Pentecostal festivities. Another response to later earthquakes in the area was to stage a mass religious procession, while in Paris, public prayers were common until 1709. A French earthquake in 1618 became a sign of God's wrath toward Protestants.[16] Conversely, an English earthquake in 1580 was a warning from God to hold fast to Protestantism and avoid relapsing into Catholicism. Responses to the 1580 event included abandoning Corpus Christi celebrations in Coventry, the imposition of a national Day of Atonement, and the addition of special earthquake liturgies. A 1601 Swiss earthquake caused the Lucerne City government to require the participation of all residences in a forty-hour devotional prayer relay from one church to the next.[17] Similar to classical Chinese notions, during the fifteenth and sixteenth centuries earthquakes in Europe sometimes became associated with political danger. "Large earthquakes," said one French commentator, "foretell war, epidemics, or tyranny." In 1660, however, for the first time a French earthquake received a positive interpretation, as "an attribute of royal power."[18] Increasingly, earthquakes became social and historical phenomena. As Gregory Quenet points out, "Earthquakes are never understood in themselves but are seen in the social and cultural context that preserves their memory, in the physical landscape that keeps their traces, in societies' institutional framework, in the communication system that spread the information, or even in the theories which were used to interpret the earthquake."[19] This book explores all of these dimensions of earthquakes in the context of early modern Japan.

In Europe during the eighteenth century, earthquakes became objects of formal study, which corresponded with the emergence of secular approaches to nature. Of particular importance was the massively destructive Lisbon earthquake of November 1, 1755, felt throughout large areas of Europe. Although talk of divine wrath remained common, much of the discourse following the earthquake revealed a "hiatus between God and nature."[20] Explanations independent of Providence emerged, such as Immanuel Kant's theory that winds passing through large subterranean caverns of hot gas caused the disaster.[21] Not only did Lisbon accelerate the emergence of scientific explanations of earthquakes, it also sparked hope that the human capacities for knowledge acquisition might enable future generations to anticipate or otherwise deal with earthquakes. As one French essay competition entrant wrote, "It would, perhaps, not be impossible to discover some sort of sign, which would indicate the coming of an earthquake, but it is not in this century that we will be able to enjoy this discovery." Significantly, this writer laid out a detailed plan for the observation and cataloguing of these signs, an approach that continues even in the present day, at least among many who regard earthquake prediction as possible:

> For this reason we must make sure, to achieve more certainty, that in each country people spend their time, day and night, examining with continuous, serious mediated and reflected attention, the differences right down to the little changes that happen to all the elements, following step by step, so to speak, their different variations, of which they will take and keep exact notes; that these observers will be replaced after their death by others, who will have worked on the same observations; and thus successively until we have experienced several earthquakes; and then bringing together all these observations, we will see whether the circumstances proceeding each quake are the same, and then we will be able to discover signs that could signal the approach of an earthquake.[22]

Precisely the same dream of future prediction, conducted in the same manner, arose in Japan during the nineteenth century, especially after 1855. Indeed, in this and other respects, the impact of the Ansei Edo earthquake on Japan was similar to that of the Lisbon earthquake on Western Europe.

Disasters, Vulnerability, and Time

Whether seismic waves caused by the rupturing of faults cause a disaster in any particular case depends on many circumstances beyond the obvious question of the quantity of energy released (magnitude). As Christof Mauch points out, "The shifting of tectonic plates, for instance, may not be absolutely predictable, but from a geological point of view it is 'normal.' The vast majority of these shifts go unnoticed, and no geologist would think of labeling them as a catastrophe. In other cases, nature may supply the trigger for a disaster, but whether we call a natural occurrence a catastrophe depends largely on our perception of its impact on humans."[23]

Nature produces a variety of hazards or potential hazards. When these hazards interact with human society, they might lead to a "natural disaster" or catastrophe, depending on that society's level of vulnerability. Vulnerability is a function of many variables. Consider one aspect of location, for example. Estimates are that a magnitude (M) 7.5 earthquake in Los Angeles would produce roughly fifty thousand victims, but this figure would increase to one million should an earthquake of the same magnitude strike Teheran.[24] The differing vulnerabilities of Los Angeles and Teheran are the result of multiple factors, the most important of which is the quality of building and infrastructure construction. Vulnerability is also a variable within a given society. Indeed, the emphasis on studies of disasters since the 1980s has been to focus on factors such as social class, poverty, or other attributes of certain social groups that result in greater vulnerability for them.[25] The limited English-language literature dealing with the Ansei Edo earthquake has assumed that Edo's less affluent residents were more vulnerable in this event. We will see, however, that in many respects this assumption is inaccurate.

Major factors contributing to a society's vulnerability to natural hazards include location, infrastructure, sociopolitical structure, economic circumstances, technology, religion, and ideology. All of these factors develop and persist over time. In other words, time is a fundamental question in the context of vulnerability. As Anthony Oliver-Smith explains, "The life history of a disaster begins prior to the appearance of a specific event-focused agent. Indeed, in certain circumstances disasters become part of the profile of any human system at its first organizational moment in a relatively fixed location or area."[26] In light of this insight, when did the Ansei Edo earthquake begin? Moreover, when did it end?

There is, of course, no "correct" answer to these questions, and one can usually push historical beginnings ever further back in time. Assuming roughly half a century for Tokugawa Japan's institutions and society to mature, and focusing on the context of major earthquakes and the emergence of typical patterns of response to them, I would propose 1662, the year of the Kanbun earthquake, as a plausible starting point. In other words, Ansei Edo was in many respects the culmination of trends that first emerged in 1662. As for the end, although Edo was largely rebuilt by roughly 1858, the Ansei Edo earthquake remained a part of popular memory and influenced understandings of earthquakes well into the twentieth century. Indeed, as I argue in this book, its legacy persists to the present. The 1995 Hanshin-Awaji (Kobe) earthquake resulted in a revival of academic interest in the Ansei Edo earthquake and in claims that a variety of precursors preceded the Kobe event—the same types of claims made in 1855 and 1856.[27] Likewise, as we have seen, the Great East Japan Earthquake and tsunami of 2011 revived interest in earthquake lore and approaches to earthquakes that emerged in 1855. It is in this sense that I refer to the Ansei Edo earthquake as having a long history and continuing legacy. Because I examine Ansei Edo and its influence in the context of major earthquakes of the early modern and modern eras, some basic information about these seismic events is necessary at the start.

The Earthquakes

Here I use the most common name for each earthquake, with major alternative names, if any, given in parentheses in the heading. The lunar date for premodern earthquakes is also given in the heading, using the format year/month/day. Because the epicenter of many of the earthquakes discussed here was located in or near the Kansai region, I begin with a brief description of the geology of this area.

Japan's Kansai (or Kinki) region, which includes the urban areas of Osaka, Kobe, and Kyoto, sits atop many active faults. It includes the so-called Kinki Triangle Region, defined by points at the edge of Wakasa Bay, Awaji

Island, and the middle of Ise Bay, which contains the highest concentration of active faults in Japan. The tectonic geology of the region is largely a result of the Philippine Sea Plate subducting under the Eurasian Plate. This subduction creates the Nankai Trough in the Pacific Ocean off the coast of the Kansai region and areas to the south. Accumulated strain from plate motion generates three types of earthquakes. The first is ocean trench earthquakes, usually originating in areas off the coast of Mie Prefecture or Wakayama Prefecture. Examples include the 1707 Hōei earthquake and the 1854 Ansei Tōkai-Nankai earthquake. Since the late 1970s, popular fear of another earthquake of this type and the tsunami it would generate has prompted significant government spending on monitoring this region.[28] Next are deep intraplate earthquakes, with focal (origin) points in the part of the Philippine Sea Plate pushing underneath the Eurasian Plate. The third type is inland earthquakes with a relatively shallow focus within the Eurasian Plate.[29]

Keichō-Fushimi (Fushimi-Momoyama), 1596 (Keichō 1)/7/13

The Keichō-Fushimi earthquake (around midnight, September 5, 1596) was a shallow inland earthquake of M7.5 with an epicenter near Hirano on the outskirts of Osaka. It caused approximately fifteen hundred deaths and extensive property damage. Efforts to map all the faults under Japan

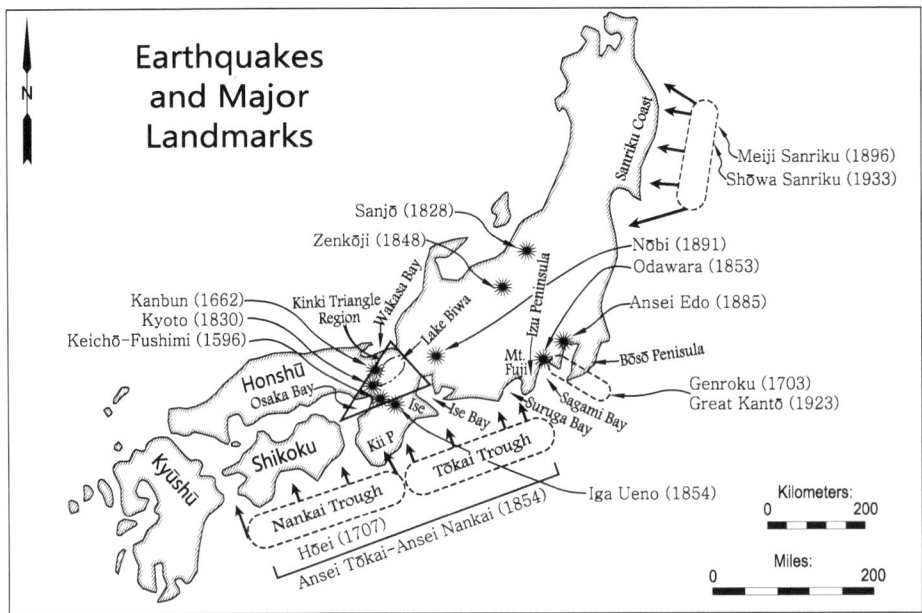

MAP 1 Produced by Jeffrey Smits, Cherokee Drafting Specialists.

after the 1995 Hanshin-Awaji (Kobe) earthquake led to the discovery of the likely cause of the Keichō-Fushimi earthquake: a rupture of the Arima-Takatsuki Fault, possibly in conjunction with the Rokkō-Awaji Island Fault.[30] The keep of Fushimi Castle collapsed, and a falling stone wall killed seventy-three high-ranking ladies-in-waiting and some five hundred maids. Toyotomi Hideyoshi, Japan's ruler at the time, was in the castle and escaped into a courtyard. Many shrines and temples in the area incurred major damage.[31]

Kanbun (Kanbun Ōmi-Wakasa; Biwakō seigan), 1662 (Kanbun 2)/5/1

Approximately ninety kilometers northwest from the epicenter of Keichō-Fushimi, another shallow inland earthquake struck around noon on June 16, 1662. The epicenter of the Kanbun earthquake was the western bank of Lake Biwa near the present-day town of Kitahama. Magnitude estimates range from 7.25 to 7.6, and people as far away as Fukuyama and Edo felt this powerful earthquake. Some seismologists posit movement of the Hiruga Fault under Wakasa Bay, followed by movement of the central and northern sections of the Hanaore Fault, which extends from a point north of Kyoto along the western shore of Lake Biwa. In this view, the Kanbun earthquake was a two-stage seismic event, and it may have been two earthquakes in succession. According to Okada Yoshimitsu, movement of the Hanaore Fault and the Lake Biwa West Bank Fault Zone caused the earthquake. Damage was severe across the region, varying significantly as a function of local conditions. Estimates of the death toll for the western shore of Lake Biwa range from seven hundred to one thousand, including about two hundred deaths in Kyoto. The situation in some places, however, was much worse than these overall figures suggest. Landslides buried several villages, wiping out the vast majority of their populations. In Kawamura in the Kutsuki Valley, for example, collapsing mountains buried all of the approximately fifty dwellings, and only thirty-seven of the more than three hundred inhabitants survived.[32]

Genroku, 1703 (Genroku 16)/11/23

The M7.9–8.2 Genroku earthquake occurred at about 2 a.m. on December 31, 1703, resulting in over ten thousand deaths. It was an ocean trough earthquake originating somewhere between Sagami Bay and the area off the southern shore of the Bōsō Peninsula. It was the same type as the Great Kantō Earthquake of 1923, one of the M8 class earthquakes whose origins are the ocean troughs caused by the Philippine Sea Plate sub-ducting beneath the Eurasian Plate and the North American Plate. These earthquakes typically are damaging in and of themselves, and they usu-ally generate tsunamis. In the case of the Genroku earthquake, tsunamis of between 2 and 10.5 meters in height struck the coast between Shimoda and Inuboesaki, washing approximately six hundred to their deaths near Kamakura. The total death toll from the seismic sea waves probably num-bered in the thousands. Shaking could be felt as far away as the Kinki region, and the earthquake and tsunami destroyed over twenty-eight thousand structures. This earthquake caused uplift in the southern part of the Bōsō Peninsula by as much as 5.5 meters.[33] Daimyō mansions on solid ground such as the Yamanote area suffered relatively light damage, whereas alluvial areas of Shitamachi suffered severe damage.[34] This earthquake prompted a change of era name from Genroku to Hōei.

Hōei, 1707 (Hōei 4)/10/4

The change of era names, however, did not mollify the cosmic forces. On October 28, 1707, nearly the entire Nankai Trough plate boundary ruptured at once. Until March 11, 2011, many specialists regarded the resulting M8.4 Hōei earthquake as the strongest seismic event to shake Japan. It caused at least five thousand deaths and possibly as many as fifteen thousand. It gen-erated tsunamis from the Izu Peninsula to Kyushu, and the waves reached heights of five to eight meters in Shikoku. They washed away nearly eighteen thousand homes and destroyed some three thousand boats and ships. Uplift and subsidence were common throughout Shikoku. Tsuro and Murozu harbors in Murodozaki (Tosa) experienced about 1.8 meters of uplift, and a twenty-square-kilometer area near Kōchi sank by about 2 meters.[35]

Sanjō (Echigo), 1828 (Bunsei 11)/11/12

At about 7 a.m. on a market day, December 18, 1828, the earth around Sanjō in Echigo (Niigata Prefecture) began to shake. The M6.9 inland, shallow-focus earthquake caused liquefaction (many buildings sank about one meter into the earth), landslides, and fissures in the earth from which water and sand flowed. The earthquake caused 1,443 deaths, collapsed 9,808 structures, and burned 1,204 more. The shaking began just after many merchants in the marketplace had started their fires to prepare food, and many fled without extinguishing them. Soon the whole town was ablaze, and the cries of trapped victims mingled with cries of those chanting the name of Amida Buddha.[36] One account mentioned the anguished sounds of horses and cows amidst smoke and people drowning in flooded pits that had opened in the ground.[37]

Kyoto, 1830 (Bunsei 13/Tenpō 1)/7/2

At roughly 4 p.m. on August 19, 1830, an inland earthquake of approximately M6.5 shook Kyoto. The epicenter was slightly to the northwest of the city. Kyoto's death toll in 1830 was between two and three hundred. Heavy shaking was limited to the city itself, although some outlying areas experienced minor damage and the earthquake was felt in nearby cities such as Osaka. Destruction of storehouses was widespread, but destruction of houses was relatively modest. Nijō Castle suffered serious damage, and liquefaction was widespread throughout the city.[38] As was usual by this time, there were reports of light flashes in the sky and bursts of light issuing forth from the earth.[39]

Zenkōji, 1847 (Kōka 4)/3/24

Disaster struck Zenkōji and its surrounding areas in present-day Nagano Prefecture around 10 p.m. on May 8, 1847, in the form of a powerful M7.4 shallow-focus inland earthquake. Activity in the Shinano Fault Zone caused the earthquake, whose epicenter was at the western part of the Nagano Basin, near Nagano City.[40] It caused massive destruction and death around the region and was particularly devastating because of the seven to eight thousand pilgrims from all parts of Japan lodged in cramped quarters around Zenkōji, a major temple. The pilgrims were there for a *kaichō*, the public display of an ordinarily hidden image of Amida and two attendants, which takes place once every seven years at the temple. According to Sangawa Akira's analysis, three to four thousand of these visitors died, many in fires, as did forty-four of the forty-six priests in attendance. The whole area inside and outside the temple was illuminated by lanterns that night, which soon led to a blazing inferno. The other dramatic locus of destruction was an area of the Sai River blocked by a landslide when Mt. Iwakura (also called Mt. Kokuzō) collapsed. Heavy rains on May 22 and 23 swelled the accumulated water. The landslide-dam gave way on the twenty-seventh, causing massive flooding downstream.[41] It was, according to one account, "a disaster of shaking, fire, and water all at once."[42] These three scourges left as many as twelve thousand dead and caused widespread destruction to fields and structures.[43] Rumors of even higher death tolls circulated in the days after the main shock. One diary entry for the twenty-ninth day, for example, includes hearsay that several tens of thousands were crushed to death by collapsing houses.[44]

Odawara, 1853 (Kaei 6)/2/2

"The earthquake of the fourth hour of the second day of this month," began one local account, "did much damage to the castle's keep. The lord's mansion suffered great damage. Nothing remained of the inner wall, the second wall, and the outer wall, which collapsed into the moat. Many pieces of the stone wall ended up in the water, and structures like the connecting watch towers were completely destroyed."[45] Similarly, a dramatic print entitled *Great Earthquake in Sagami* (*Sagami no kuni ōjishin*) showed the area to the north-northeast of the castle in flames and claimed a greatly exaggerated death toll of 3,780.[46] The more likely death toll from this M6.7–7.0 earthquake that struck around 10 a.m. on March 11, 1853, was less than one hundred. Damage to structures was severe in areas with a poor soil base, especially to the north-northeast of the castle. It appears that two main shocks occurred about ten to fifteen seconds apart, and the earthquake generated a tsunami that came ashore at Manatsuru but did little serious damage. Shaking could be felt in parts of Edo and produced some minor damage there.[47]

Iga-Ueno (Ansei Iga), 1854 (Kaei 7/Ansei 1)/6/15

Movement along the Kizugawa Fault Zone in western Mie Prefecture caused the Iga-Ueno earthquake. It was a M7.25 shallow inland earthquake that struck around 2 a.m. on July 9, 1854. Over twenty precursors shook the ground prior to the main shock. Starting on the twelfth day of the lunar month, and again on the thirteenth day, two especially large precursor earthquakes shook the area. Foreshocks combined with aftershocks constituted an ideal combination for putting the population of the area psychologically on edge. One typical account explains that people "only worried, forgetting to eat and sleep" and that "although there has never been a natural disaster in the country of Iga since the age of the deities, the frequency of daily shaking goes up and down, and just tonight there were six or seven shakes."[48] The death toll was about thirteen hundred, and the shaking destroyed nearly six thousand houses and shops. The earthquake shook Kyoto and Osaka, and it caused significant damage in and around Iga, Nara, and Yokkaichi.[49] In Osaka, many residents near the harbor or waterways fled to the safety of boats, and because this inland earthquake did not generate a tsunami, they suffered no problems. About six months later, however, during the Ansei Nankai earthquake, such a move proved fatal for several hundred tsunami victims.[50]

Ansei Tōkai and Ansei Nankai, 1854 (Kaei 7/Ansei 1)/11/4 and 11/5

At 9 a.m. on December 23, 1854, a M8.4 ocean trench earthquake occurred along the subducting edge of the Philippine Sea Plate between the Kumano-oki (offshore from the Kii Peninsula) northeast to the Enshū-oki (Suruga Bay, Shizuoka Prefecture). Shaking could be felt as far south as northern Kyushu and as far north as the Tōhoku region. Tsunamis struck areas between the Bōsō Peninsula and Kōchi Prefecture, with wave heights ranging from 4.5 to 10 meters. The most severe ground shaking occurred along the coast between Numazu and Ise Bay. In some places, damage to structures approached 100 percent. The estimated death toll from the Ansei Tōkai earthquake is two to three thousand, and the shaking and sea waves destroyed as many as thirty thousand structures.[51]

The next day, at approximately 4 p.m., an adjacent segment of the subducting edge of the Philippine Sea Plate ruptured off the eastern coast of Shikoku. This Ansei Nankai earthquake was essentially a continuation of the previous day's seismic event, and some areas such as the southern tip of the Kii Peninsula were shaken by both earthquakes. In the village of Koza, for example, residents spent the night after the Tōkai earthquake in the hills. They came back to what was left of their houses the next day and were shaken by the Nankai earthquake that afternoon. Seeing the sea recede, they escaped a second tsunami just in time. The Ansei Nankai earthquake shook a wide area, and ground motion was especially severe in Shikoku and adjacent coastal areas of Honshu. The death toll from the shaking and seismic sea waves amounted to several thousand. Ocean trench earthquakes often cause uplifting and subsidence, and this one was no exception. Major earthquakes during the Hōei, Ansei, and Shōwa eras raised the port of Murozu in Kōchi 1.4, 1.2, and 1.1 meters respectively. Other areas of Shikoku sank. The Tōkai earthquake caused uplift along the coast of present-day Shizuoka Prefecture, permitting construction of the Satta Pass to reconnect the Tōkaidō highway, which the earthquake had blocked.[52] Today, the Tōmei Expressway and Tōkaidō honsen rail line both pass through this uplifted area.[53]

Ansei Edo (Ansei Earthquake; Great Ansei Earthquake), 1855 (Ansei 2)/10/2

The Ansei Edo earthquake struck at approximately 10 p.m. on November 11, 1855. Estimates of magnitude vary slightly, but 6.9–7.0 is typical. Using the JMA seismic intensity scale (*shindo*), most areas of Edo were a strong or weak five. For people, a strong five would include feelings of extreme fear, the shaking impeding their movement. Modern wood-frame structures with low resistance to shaking would likely suffer damaged walls and support posts or become tilted. Generally, wooden structures in 1855 were significantly less resistant to earthquake shaking than modern buildings. Some key areas of the city suffered damage at a level of six. A weak six level of seismic intensity is characterized by difficulty standing, which would become impossible in a strong six. For wooden structures, those with poor resistance to shaking might collapse in weak six conditions, and most would collapse in strong six conditions. As a rough guide, the percentage of destruction to structures would be less than 1.5 in weak five conditions and would approach 70 in strong six conditions.[54] A relatively modest difference of location on the night of the earthquake would have determined the difference between a rude awakening and minor damage to one's residence versus complete collapse of major structures and hundreds of people being crushed. Honjo (Kuroda-ku), Fukagawa (Kōtō-ku), "Daimyo Lane" (the area roughly between Yūryakuchō and Tōkyō Stations), Asakusa, and nearby areas experienced shaking in the six range, as did Shin-Yoshiwara. Storehouses suffered extensive damage throughout the city, even within areas in the five range.[55] Deaths for civilian and military personnel combined were in the range of eight to ten thousand.

The Ansei Edo earthquake was an inland earthquake that caused choppy coastal waters but did not generate a tsunami. Estimates of the epicenter consistently place it between the northern part of Tokyo Bay and Etō-ku, at approximately the mouth of the Arakawa River as it empties into Tokyo Bay. The question of depth is less clear. Some estimates point to a shallow focus, while other studies have suggested a medium depth. There are nine extant accounts in historical materials that mention in some way the time difference between shaking caused by P-waves (pressure waves) and S-waves (shear waves). Seismologists have attempted to estimate the focus in part by examining these materials, but they are far from precise. The question of depth and precise location of the hypocenter remains open.[56]

Nōbi (Mino-Owari), 1891

At 6:38 a.m. on October 28, the Neodani Fault in northern Gifu Prefecture ruptured. The force generated by the strike-slip earthquake pushed one edge of the fault approximately six meters upward, and the fault scarp remains visible at Midori, in Motosu City, Gifu Prefecture (fig. 1).

Seismic waves extended outward from the Neodani Valley, causing extensive destruction and loss of life in Gifu and Aichi prefectures and the cities of Gifu and Nagoya. The M8.0 earthquake killed 7,273, injured 17,175, destroyed approximately 140,000 structures, and damaged another 80,000. The Nōbi earthquake was felt from Sendai in the north to Kagoshima in the south. Areas of most severe ground motion reached seven on the Japanese seismic intensity scale, the maximum value.[57]

Meiji Sanriku, 1896

The Tōhoku region is subject to shaking from fault ruptures in four types of locations: (1) ocean trench earthquakes off the Sanriku or Fukushima

FIGURE 1 The Neodani fault scarp in 2011. The 1891 Nōbi earthquake raised the landmass to the right of the fault approximately six meters. Photo by Gregory Smits.

coast; (2) intraplate earthquakes originating under northern Honshu or the Pacific; (3) shallow-focus inland earthquakes; and (4) intraplate earthquakes originating under the Sea of Japan. Three of these four types of earthquakes often generate tsunamis.[58] The Meiji Sanriku earthquake was an example of the first type.[59]

Along the Sanriku coast, the geographical interface between human settlements and the sea has often amplified the destructive power of tsunami waves. The coastline is deeply indented, and many of its bays and inlets are shaped roughly like a bugle, the bell end of which opens to the ocean. Fishing villages have been located at the narrow interiors of these bays at the precise point where tsunami wave height would be highest when a fixed quantity of water is forced into an increasingly narrow, shallow area.[60] The maximum wave height at Ryōri Village was 38.2 meters. This situation is one reason that the tsunami from the Meiji Sanriku earthquake killed twenty-two thousand when it struck on the evening of June 15. It was the deadliest tsunami in Japan's history, and it is likely to remain so even after the March 11, 2011, disaster in the same area.[61]

One important phenomenon that occurs in perhaps 10 percent of seismic events in the Tōhoku region is that an earthquake causing only mild ground motion produces a large, destructive tsunami. These dangerous events are known as "tsunami earthquakes," a term coined in 1972 by Hiroo Kanamori of the California Institute of Technology. The Meiji Sanriku earthquake and tsunami of 1896 is the most recent example of the earth shaking so mildly that people did not expect the massive tsunami wave trains that followed. Two causal factors responsible for making Meiji Sanriku into a tsunami earthquake were the fault rupturing slowly and the rupture breaking and displacing accumulated sediments at the plate boundary.[62] The best estimate for the Meiji Sanriku earthquake is that a portion of the fault about 250 kilometers long ruptured over the course of about one hundred seconds. The resulting seismic land waves would have had such a long period that people would barely have felt them. The relatively slow fault rupture pushed unconsolidated sediments upward to produce a massive tsunami.[63]

Great Kantō, 1923

At two minutes before noon, while many cooking fires were burning on a windy day, a M7.9 earthquake shook Tokyo, its suburbs, and nearby areas of Kanagawa Prefecture. The Great Kantō Earthquake of September 1, 1923, remains the deadliest seismic event in Japan's recorded history, killing over

105,000. The epicenter was located in the northwest part of Sagami Bay. Although many people died as a direct result of the shaking, more deadly were the firestorms that swept through this densely populated urban area. The earthquake also generated tsunami waves, with maximum heights ranging from six to twelve meters throughout the area. Uplift in the region was common, reaching a maximum of about two meters.[64]

Shōwa Sanriku (Sanriku), 1933

At 2:31 in the morning of March 2, 1933, the M8.1 Shōwa Sanriku earthquake shook residents awake. Originating near the location of the 1896 Meiji Sanriku earthquake, the 1933 earthquake was the result of the rupture of an intraplate fault. Although geologically different from its 1896 predecessor, the 1933 earthquake was functionally similar in that it threw off large tsunami waves. The maximum wave height at Ryōri was 28.7 meters. The Shōwa Sanriku earthquake and tsunami resulted in slightly more than three thousand deaths and over one thousand injuries. The relatively low death toll compared with 1896 was in part the result of severe ground motion from the earthquake and social memory of the 1896 tsunami prompting people to flee to high ground. The shaking, waves, and fire combined destroyed over six thousand houses, and local residents endured seventy-seven aftershocks of M6 or higher for six months following the main shock.[65]

Early Modern Japan

Because the majority of this book deals with Japan's early modern era, some background concerning relevant features of society is necessary for making sense of later discussion. The following material does not attempt systematically to describe early modern Japanese society. Instead, the focus is on points that contextualize the analysis in subsequent chapters.

Social and Political Geography

Tokugawa Japan was a patchwork of semiautonomous territories governed by some 250 warlords, commonly known as daimyō (literally "big names"). Interspersed among these daimyō domains were parcels of territory belonging to the shogun (*shōgun*, "general"), whose capital was the city of Edo, modern Tokyo. The government of the shogun (bakufu, shogunate) controlled approximately one-fifth of the land of the Japanese islands, including major cities and mines. The bakufu also conducted foreign relations, typically with the assistance of strategically located daimyō. Although daimyō enjoyed a high degree of autonomy, they were subordinate to the shogun. One manifestation of this subordinate relationship is that daimyō spent half of their time resident in Edo, technically in attendance on the shogun. Therefore, daimyō maintained mansions in Edo and in their home domains, the main reason that until the 1860s approximately half of the population of Edo was military personnel (samurai). The modern term for daimyō domains is *han,* and the polity consisting of the bakufu and daimyō domains is often called the *bakuhan* state.

Some larger daimyō domains included parcels of territory long governed by locally powerful families. In such cases, a relationship obtained between these families and their daimyō overlord similar to that obtaining between daimyō and the shogun. In internal discourse, many daimyō domains and even some sub-daimyō territories referred to themselves as *kuni* (*-koku* in compound words), a term that now means "country" in the sense of a national state.[66] Today, there is only one *kuni* in the Japanese islands, but in early modern times, depending on the context, there might be many. It is in part for this reason that Mark Ravina has characterized Tokugawa Japan as a "compound state."[67] Early modern Japan consisted of countries within countries, with the shogun's government either directly administering or exerting hegemony over the largest of these countries, Japan (Nihon, Nippon).

Militarily, the shogun's government was the most powerful political entity in early modern Japan, but its roots in the symbolic realm were shallow. When early modern Japanese spoke of Japan as a whole, they often imagined a political geography based on the classic sixty-six provinces (occasionally sixty-eight), with the imperial court in Kyoto at its center. Similar to larger daimyō domains, these provinces were also known as *kuni* (*-koku* in compound words). Although individual emperors tended to be obscure and possessed very little personal political power during most of the Tokugawa period, the imperial institution had deep roots in the symbolic and temporal fabric of Japan. The prestige of the imperial court was high in many academic circles, especially during the latter half of the Tokugawa period. Popular perceptions of the imperial court tended to be vague, but the Ise Shrine complex became a major focus of popular mass pilgrimages during the latter half of the Tokugawa period. The inner shrine at Ise housed Amaterasu, the solar deity from whom the imperial family claimed descent.

In part because they shook wide areas, major earthquakes transcended prevailing political boundaries and often served as contexts for speaking of Japan as a whole or for speaking of large portions of Japan that transcended domain boundaries. The most common way of doing so was to employ the geographic vocabulary of the provinces and imperial court. For example, a work produced after the Sanjō earthquake begins by describing the geography of "our Echigo country" (Echigo Province).[68] Writing about the Zenkōji earthquake, a local scholar's discussion ranged seamlessly from "our Shinano country" (Shinano Province) to Japan as a whole, which he called "our country" and characterized as "a land of deities superior to all others, whose people possess superior wisdom." Moreover, in Japan, "the five grains flourish, and we receive divine favor."[69] These excerpts are typical examples of early modern discourse in which the boundaries of "our country" could expand or contract depending on the context. When discussing Japan as a whole, religious imagery was common. At least according to early modern earthquake literature, "Japan" was largely a religious construct, more likely to be called *shinkoku* than Nihon. *Shinkoku*, which we might tentatively translate as "land of deities," is a term with a long and complex history. I examine it further in later chapters, but the point here is simply that earthquakes often brought a religious image of Japan to the fore.[70] Moreover, this deity-filled land was the domain of a human emperor, whose court, *honchō* (our court), was another common metonym for Japan.

There was, of course, a pragmatic reason for a geography of imperial provinces when speaking of earthquakes. The classical boundaries of Shinano, for example, encompassed several daimyō domains and several parcels of bakufu land in early modern times.[71] It was simply more convenient to describe the interactions of earthquakes and geography in the relatively broad terms of provinces. Moreover, the provinces frequently appeared in other forms of popular discourse, especially the many guidebooks that circulated widely from the late seventeenth century onward.[72]

Status categories were the basis of social organization in Tokugawa Japan. The precise number of these categories and their boundaries differed from government to government and with local circumstances. A category such as courtesan, for example, would only have applied in urban areas with licensed quarters. Most jurisdictions recognized the categories of aristocrat, cleric, warrior (samurai), townsperson, farmer or peasant, fisher person, and outcast. Writing around 1816, Buyō Inshi, a warrior from Edo, organized his account of society largely along the lines of status categories: warriors, peasants, clerics, physicians, blind people, townspeople, prostitutes, actors, and outcasts.[73] Formal status categories corresponded to the occupation of a particular household, but individuals might pursue other ways of making a living. A person born into a well-to-do farming household would be a peasant in terms of status. If he received a good education and his labor was not needed on the farm, he might make a living as, for example, a teacher. Intellectuals and artists were not status categories, yet academic and artistic activity flourished during the Tokugawa period because people from a variety of status groups pursued these activities as either amateurs or professionals. To take one example, the Confucian scholar Nakae Tōju (1608–1648) began life as a low-ranking samurai with time on his hands for study. As a young man, he resigned his post as a warrior, settled in a rural village, made his living operating a brewery, and earned a reputation as an intellectual. The point is that there was permeability at the borders of status categories, and a person's status did not necessarily correspond to how that person made a living. Social boundaries in Tokugawa Japan were sufficiently flexible to accommodate a high degree of complexity.[74]

Particularly significant was the division between warriors, comprising about 7 percent of Japan's total population and roughly half of Edo's population, and commoners (townspeople, fisher people, and farmers), who comprised most of the rest of the population. Broad categories like "commoners" or "townspeople" encompassed a wide range of possibilities, especially in large urban areas. For example, a 1687 book of lists, *Dappled Fabric of Edo (Edo kanoko)*, mentions roughly three hundred master artists in over forty categories, seven hundred master craftsmen and merchants in almost two hundred categories, and twenty varieties of wholesalers.[75] A similar diversity of crafts and trades flourished in Kyoto and other urban areas, and the cities became larger and more diversified over time. By the nineteenth century, "commoners" in Edo ranged from unskilled manual laborers newly arrived from rural areas to merchants of such wealth and power that some managed the finances of daimyō domains. Heads of villages and the city elders and neighborhood heads of Edo were de facto government officials. They sometimes wore swords in the manner of samurai when operating in an official capacity, yet they, too, were commoners.

Some earthquakes brought social tensions to the surface that might reasonably be called "class" divisions, even though all involved were commoners in a narrow legal sense of status categories (townspeople in the cities or farmers in rural areas). For example, *Outward-Bound Ship of the Wealthy*

(*Marumochi kara no defune*) is the title of a catfish print (*namazue*), one of the hundreds that appeared in the wake of the Ansei Edo earthquake. In the print, a catfish representing the earthquake forces a rich merchant atop a high point to vomit gold coins. Skilled laborers below scramble to scoop up the coins. In the text of the print, the wealthy man says that if only he had known he would lose his money so quickly he would have spent it earlier. The unsympathetic catfish points out that his actions have caused ordinary people much grief. The workers on the receiving end, citing the unexpected nature of the earthquake and the ephemeral nature of money, declare that they will spend their windfall profits at the temporary brothels, authorized after the earthquake destroyed the main licensed quarters.[76]

This print is one example of an earthquake functioning to redistribute wealth, in this case from prosperous merchants to skilled laborers, two varieties of townspeople. An implied point is that one should not hoard large quantities of money. Indeed, withholding large quantities of money from circulation and similar acts such as hoarding commodities to drive up prices were major complaints against elite merchants that sometimes resulted in violence. In the typical social theory of the day, the proper function of merchants was to circulate goods and money. Blockages in this process cause social imbalances, just as blockage of the flow of vital fluids in the body causes disease and blockage of yang energy within the earth causes earthquakes.[77] I examine these matters in detail in later chapters, but the main point here is to highlight that de facto divisions in social class based on wealth and modes of making a living did not necessarily correspond to the boundaries of formal status categories.

Bakufu Power

Headed by a shogun and administered at the top by *fudai* daimyō (lords of privileged domains), the bakufu was primarily a military organization concerned with the regulation of Japan's warriors. One aspect of the status-based society described above is that most status groups were semiautonomous. The bakufu periodically issued directives regarding civilian affairs, but in most routine matters it expected civilians to regulate themselves. The major bakufu concerns with respect to civilians were that peasants in bakufu territory pay their taxes and that peasants and townspeople everywhere refrain from collective violence or serious social disruption. By the eighteenth century, as a complex market economy connected cities and countryside, the bakufu became increasingly concerned with regulating economic affairs connected with the supply and prices of goods.[78] Although it sometimes issued regulations about cultural or moral matters, the shogunate lacked the will and the means to enforce most such regulations for more than a brief period. Buyō Inshi pointed out in the early nineteenth century that "the Shogunate's proclamations and ordinances are called 'three-day laws.' No one fears them, and no one pays attention to them. . . . They are disregarded after that short period of time."[79] Because punishment for violating bakufu edicts was rare and not usually severe by modern standards, "disregard of regulations in the Tokugawa period was the norm."[80]

One reason people so commonly disregarded bakufu decrees in the capital was because the shogunate devoted few resources to civil administration and police. To ensure at least a modicum of civil order, police officials tended to rely on the cooperation of neighborhood-level leaders, trade guilds, and other commoner organizations. As historian Katō Takashi points out, in Edo, "the veneer of administration over commoner residential areas was thin: the population was great, the numbers of administrators and police relatively small. Consequently, to ensure that merchant and artisan quarters of the city ran in an orderly and peaceful fashion, the city magistrates delegated many important functions to the merchants and artisans themselves."[81] To point out these systematic limitations in direct bakufu control of the civilian population is not to suggest that the shogunate was completely incapable of enforcing its decrees. In the wake of the Ansei Edo earthquake, for example, the bakufu eventually regained control over publishers, forcing them to stop producing catfish prints and other unauthorized items. Getting to this point, however, required two months of time and several rounds of escalation, culminating in the brief arrest of the heads of several publishing guilds.

When bakufu officials focused their limited resources, they could often force compliance with most unpopular decrees, at least temporarily. Even when there was a will to enforce bakufu edicts, however, sometimes opposing forces were simply too strong. Bakufu attempts to control prices and wages after the Ansei Edo earthquake, for example, were a complete failure despite the repeated issuing of sternly worded edicts and some concrete attempts to enforce them. The power of supply and demand was simply too strong and pervasive. Overall, it is important to bear in mind that there was usually a substantial gap between ideal behavior as framed by bakufu directives and the actual circumstances of life among peasants and, especially, townspeople.

Protest and World Renewal

Tokugawa Japan was a litigious society, with serious disputes over the control of wealth usually taking the form of petitions to higher authorities.[82] Occasionally, the usual social mechanisms could not contain disputes, and they became violent. Usually this violence was directed against property, and it conformed to customary patterns. Of the possible grievances that might cause peasants or townspeople to protest by destroying the property of wealthy farmers or merchants, the high price of rice, allegedly caused by hoarding, was perhaps most common.

The following dialogue between an accused peasant rioter in 1836 and a government official is typical of violent protest in several respects:

> *Tatsuzō:* We got together at Ishimidō to put on a "festival of righteous world revival." We wanted to bring relief to those who were suffering so much.
>
> *Interrogating officer:* What kind of nonsense is that—a "festival of righteous world revival"! You break into the households of respected merchants; smash apart casks of sake. You call that a festival to rectify the world? You have the audacity to claim that your actions stand "apart from the law"?
>
> *Tatsuzō:* To hoard rice; to take this rice that sustains us in this transient life and squander it on making sake—that is what causes suffering for so many people. . . . We got together in order to make an appeal to this respected merchant, to beg him for some food. None of us ever intended to destroy his shop and home. It just happened that we got into a quarrel with him and a fight broke out.[83]

Tatsuzō was attempting to evade punishment by claiming that his group was engaging in a peaceful protest against a genuine injustice that happened, unintentionally, to devolve into a private quarrel. Tokugawa period governments generally regarded nonfatal quarrels as private matters, outside the bounds of the legal system. Regardless of its precise legal status, the smashing of the liquor casks in this case was a fight between two groups of rural commoners. Protests and riots in Tokugawa Japan usually resulted from disputes between groups of commoners, even if warriors became involved to restore order.

The term "world renewal" in the above dialogue is significant. The Japanese terms would be *yonaoshi* and *yonaori,* with *yo* literally meaning "world" but almost always referring to the immediate local society and *naoshi/naori* meaning rectification, correction, or renewal. Especially in the latter decades of the Tokugawa period, it was common to frame public protests as instances of social renewal or rectification, which implies that the protestors are on the correct side of righteousness and the cosmic forces. Indeed, during the last half of the Tokugawa period it became common for peasant protesters to claim that they were acting as agents of deities with names like Yonaoshi Daimyōjin or Yonaoshi Kami, thus framing any destruction or disruption they might cause as an act of divine retribution.[84] Significantly, however, protests or riots conducted in the name of world renewal were rarely revolutionary in their goals. As Herbert Bix points out, "Most often *yonaoshi* denoted a world-affirming experience. One engaged in such actions to exorcise the evils of local society, thereby preventing the world from coming to an end."[85]

The terms *yonaoshi* and *yonaori* were also associated with earthquakes during the Tokugawa period. It was not until 1855, however, that the meaning of earthquake-related *yonaoshi* merged with that of social renewal in the context of protests by peasants or townspeople. In the context of pre-1855 earthquakes, "*Yonaoshi, yonaoshi!*" was a talismanic chant with roots in shamanic purification rites. *Yonaori,* a word derived from the intransitive form of the verb, functioned the same way. It was something that people would have said or shouted after the earth began to shake, similar to shouting "*Kuwabara, kuwabara!*" at times of severe thunder and lightning. "*Yonaoshi*" was a common chant in the Kansai area, as Asai Ryōi's description of the main shock of the 1662 Kanbun earthquake indicates: "When people realized it was an earthquake they shouted '*Yonaoshi, yonaoshi!*' and houses large and small began to sway and shake."[86] A portrayal of the 1830 Kyoto earthquake similarly describes everyone chanting "*Yonaoshi!*" when the shaking started.[87] Chanting "*Manzairaku!*" was more common in the Kantō area (the vicinity of Edo) during earthquakes. After the major earthquakes of 1853 and 1854, *yonaoshi* began to appear in the titles of prints and in other written materials with great frequency, but it retained its meaning of a talismanic chant. One significant feature of the Ansei Edo earthquake is that Edo's townspeople interpreted the event itself as an instance of *yonaoshi,* the first time an earthquake rhetorically played such a role.[88]

Mass Media and Literacy

Mass media played a major role in shaping Tokugawa society. The early bakufu looked with suspicion on popular publishing, mainly because of its potential to disrupt society by spreading rumors. During the seventeenth century, the bakufu was able to police the spread of rumors to some extent, but during the eighteenth century urban population growth and the development of informal news networks that linked major cities facilitated the rapid spread of sensational news. Moreover, starting in the late eighteenth century, the demand for ostensibly objective information began to transcend class and status groups. Certain types of news, especially natural catastrophes, were of interest to the entire society. In this context, information became an economically valuable commodity.[89] The bakufu attempted in various ways to restrict or regulate the public dissemination of information. For example, an edict prohibiting "the publication of popular ballads or items concerning 'strange events that have happened recently'" and threatening serious punishments for violators appeared in 1684. Significantly, this edict was reissued in 1698, 1703, and 1713, an indication that it was difficult or impossible for the bakufu to suppress information with popular appeal.[90] "Censorship was not a joke in the Tokugawa period," concludes Peter Kornicki, "but neither was it applied so harshly or consistently as to shackle authors and publishers and force them to publish works of the kind that were acceptable to the Bakufu."[91]

Sometimes the bakufu tried to leverage news networks by leaking or otherwise providing information to publishers. As early as the 1640s, for example, the bakufu cooperated with the commercial publishing of military mirrors, a genre of guidebooks that listed extensive information about bakufu and domain officials.[92] In the weeks after the Ansei Edo earthquake, detailed casualty statistics compiled by the City Magistrate's Office appeared in the popular press. Moreover, by publishing the names and deeds of businesses and individuals who received bakufu rewards for their contributions to the relief effort, the press helped multiply the impact of government aid. In other words, despite general distrust of the popular press, entities of the bakufu were willing at times to work with or leverage mass media.

One important issue connected with the spread of written information is literacy. Literacy is difficult to define and measure, but nearly all studies agree that literacy rates increased throughout the Tokugawa period and that they were substantially higher in urban areas compared with the countryside. Schooling of various kinds was one important contributor to literacy rates. Shimizu Isao posits that the availability of inexpensive popular prints itself encouraged basic literacy. In any case, prints produced early in the Tokugawa period featured very little written text. In nineteenth-century prints, by contrast, the spaces between the illustrations were usually packed with text. In other words, literacy among townspeople was sufficiently high that prints with a high density of text became commercially viable.[93] In a thorough study of literacy throughout the Tokugawa period, Richard Rubinger concludes that "by the nineteenth century a relatively high percentage of the urban and working population may have had at least *kana* literacy," referring to the *kana* syllabary that functioned like a basic alphabet.[94] Similarly, Kornicki concludes that urban literacy rates in the late Tokugawa period were high, in part because literacy was valuable "for reasons of employment, leisure, and dealings with the authorities."[95] One other relevant point is that literate people were not the only consumers of the printed word because "reading aloud to groups helped overcome low rates of literacy," a phenomenon that would have been especially common in densely populated urban areas.[96] Major earthquakes in early modern Japan produced large quantities of written materials, ranging from simple broadside prints to academic books. The content of much of this material reached a wide audience, either directly or indirectly.

Religious and Intellectual Milieu

Making sense of earthquakes in Tokugawa Japan took place within an intellectual milieu ranging from popular religious practices and beliefs to scientific thought. Both popular and academic discourse in Tokugawa Japan was diverse, and "the Japanese," or any significant subset of them, held a broad range of views and opinions. The stress of earthquakes tended to highlight some of these views. Subsequent chapters will explore particular areas in detail, and the basic outline here serves as a point of departure.

The religious environment in medieval Japan included many forms of Buddhism, most of which were linked with local Japanese deities, the *kami*. A common Buddhist worldview in the medieval era saw Japan as a small, remote, peripheral land. A common metaphor was that the Japanese islands were like foam floating in the sea. Not only was Japan geographically remote, but temporally it had entered a degenerate phase of the Buddhist cycle (*mappō*), which made the attainment of enlightenment difficult or impossible. Therefore, the buddhas took pity on Japan and manifested themselves as *kami* to guide people toward salvation. For this reason, Japan was a *shinkoku*. In many, but not all, medieval contexts, *shinkoku* was a term indicating inferiority in the sense that Japan needed and received special treatment by the buddhas. China, by contrast, was the "sagely country" (*seikoku*) because the buddhas had manifested themselves there as sages like Confucius and Laozi.[97] In addition to buddhas and *kami,* the religious world of medieval Japan included Chinese-derived deities such as Enma, one of the judges or "kings" of hades, and other powerful entities such as the stars of the Big Dipper that were neither *kami* nor buddhas or bodhisattvas.

This complex religious mix carried over into early modern Japan, albeit with changes in emphasis, and it became even more diverse. Combinations of Buddhism, *kami* belief, other deities or powerful entities, and sometimes Confucian-derived ethical codes mixed and merged, especially during the nineteenth century. Noguchi Takehiko likens the religious environment in late Tokugawa Japan to the radio broadcasting spectrum. One needed only slowly to turn the dial to tune in to one religion after another.[98] Moreover, certain deities became brief stars on the cosmic stage. These *hayarigami* (rapidly popular deities) tended to attract popular attention quickly, only to fade away with equal speed.[99] As the rebuilding began after the Ansei Edo earthquake, even the earthquake catfish assumed its brief place as a deity-of-the-moment.[100]

Serious earthquakes caused concern about the viability of society's foundations. This concern was one reason earthquakes often resulted in literature discussing their causal mechanisms and the history of such events in Japan and beyond. It was not the case, however, that the bakufu or domain governments attempted to suppress reporting on earthquakes or other natural disasters out of fear that these events implied criticism of the political order.[101] For one thing, such suppression would have been impossible. Furthermore, many earthquakes shook territory that encompassed several different political entities, thus weakening the sense of cosmic forces striking a particular territory.

Earthquakes whose damage was limited to places of great political significance such as Kyoto in 1830 or Edo in 1703 and 1855, however, did generate considerable anxiety precisely by virtue of location. The Genroku and Kyoto earthquakes, for example, caused the era name to change.[102] Matsuzaki Kōdō, a Confucian scholar living near Edo, took anxious notice of both the 1830 Kyoto earthquake and the unseasonable blooming of cherry trees. Writing in his diary a day after the Tempō era started, he said,

"Our ruler is virtuous, and our habits upright . . . so there should be no reason for any disasters. . . . All we can do is pray for the Heavenly Protection of yesterday's new era name."[103] One common manifestation of anxiety about society in the wake of major earthquakes was a reaffirmation of the ruler's virtue, even if it was not always clear whether that ruler was the emperor, shogun, or one or more daimyō. I examine other responses to this anxiety in later chapters. With the Ansei Edo earthquake as a partial exception in the realm of rhetoric, major earthquakes in early modern Japan did not produce calls for a change of government, much less action to that effect.

The classic Chinese idea of the Mandate of Heaven regarded phenomena such as epidemics, crop failures, and natural disasters as warnings that human society and heavenly principles were out of alignment. Typically, the fault for this situation lay with the ruler, whose identity was usually obvious in China but not necessarily so in early modern Japan. This Chinese idea was well known and broadly accepted in Japan. When Edo shook in 1855, for example, prominent bakufu official Matsudaira Shungaku reacted in part by writing a memo to Abe Masahiro, the de facto head of the shogunate. Shungaku listed recent earthquakes, other natural disasters, and the unwelcome visits of American, Russian, and British naval vessels. Together with the present disaster in Edo, these events "definitely constitute a heavenly warning," he concluded.[104] Compared with Japan's medieval era, however, the notion of cosmic retribution was not a conspicuous component of early modern political thought. Moreover, some prominent intellectuals such as Yamaga Sokō (1622–1685), Miura Baien (1723–1789), and Motoori Norinaga (1730–1801) explicitly rejected the idea of the Mandate of Heaven or argued that it did not apply to Japan.[105] In short, there was a range of views among elite Japanese regarding possible links between human society and cosmic forces.

At the level of popular discourse, a rich rhetorical palate of symbols characterizing misalignments between human society and cosmic forces had developed by the nineteenth century. In the context of discussing popular broadsides (kawaraban) at this time, Gerald Groemer has this to say:

> Kawaraban hermeneutics could rely on a complex system of signs, symbols, analogies, correspondences, and metaphors that existed in the context of everyday life and effortlessly crossed the fluid borders of science, magic, astrology, folk belief, political/moral ideology, literature, poetry, and religion. This context of interpretation allowed

explicators with sufficient insight and imagination to apprehend cryptic messages of Heaven, and to endow the seemingly accidental with a meaning and causal necessity that spoke directly to the concerns of reader or listener. Conveniently enough, Heaven often communicated through newsworthy events. Sudden and disastrous natural phenomena could signal blunders of an inept government that had set nature out of balance with society. Similarly, large-scale social phenomena such as fads, crazes, and rumors might also be interpreted as a portentous sign.[106]

The potential for earthquakes to be interpreted as meaningful cosmic events undoubtedly added to survivor anxiety soon after the main shock, particularly amidst aftershocks. One could not be sure what causal forces were at work and whether the worst of the shaking, or other dangers, had passed. After the earth calmed, however, the tendency to connect natural disasters and social problems lacked sufficient focus and sustenance to undermine political legitimacy. We will see, however, that one line of interpretation after the Ansei Edo earthquake did anticipate major changes in society similar to those that actually occurred twelve years later.

Nearly all early modern Japanese understood the mechanical functioning of the world and the cosmos in terms of balances and interactions between yin and yang. The popular press, for example, typically characterized earthquakes as the result of imbalances in yin and yang. One explanation of earthquakes in a two-sheet print describing the Ansei Tōkai/Nankai earthquake of 1854 begins, "When cold and hot, warm and cool circulate normally, all is well and there are no abnormalities such as earthquakes and thunder." However, "When yin wells up, pressing against yang, earthquakes of varying intensities occur as a function of the strength of that pressure." The passage then goes on to explain specific effects of earthquakes such as the upwelling of new springs or "fire energy" issuing from fissures in the ground.[107] Early modern Japanese not only explained earthquakes and thunder (often regarded as the same basic phenomenon) in terms of yin and yang, but most other aspects of the natural world were similarly accounted for. Even intellectuals who found classical or recent European ideas intriguing often transposed them into a framework based on yin and yang.

Yin-yang was a useful theoretical construct, in part because it could encompass a wide range of phenomena. It served as the starting point for complex systems of correlative cosmology and thus possessed great explanatory power. Most scientists in the Meiji era quickly abandoned yin-yang, relegating it to a quasi-religious realm that many modern people came to regard as superstitious. Nevertheless, it served for centuries as the dominant academic paradigm for apprehending the natural world. With respect to earthquakes, yin-yang coexisted with, and by the eighteenth century overshadowed, Buddhist cosmology. Compared with Buddhist cosmology, yin-yang could more elegantly explain main shocks, aftershocks, liquefaction, sand blows, uplift, and other observable phenomena associated with earthquakes. As we will see, however, the Ansei Edo earthquake should never have occurred according to the prevailing yin-yang–based theory of earthquakes. The academic community, therefore, became open to new theories about earthquakes after 1855. Nevertheless, remnants of yin-yang thinking about earthquakes persisted well into the modern era in both the popular imagination and in academic circles. Indeed, many of the alleged precursors of earthquakes advanced by contemporary scientists advocating the possibility of earthquake prediction are remarkably similar to early modern theories of heat or other forms of yang energy moving the earth.[108]

Catfish

A book on the impact of earthquakes in human history claims that records of earthquakes in Japan "have been kept faithfully since 481 CE, a time when seismic events were thought to be caused by the wriggling of a gigantic catfish that lived in the sea beneath Japan and supported the islands on its back."[109] Writing in an art magazine, Hidemi Shiga says, "In premodern Japan, people believed that namazu (catfish) living under the earth caused earthquakes."[110] Seismologist Tsuji Yoshinobu, in a paperback book describing earthquake mechanisms for general readers, includes a one-page column, "Edo Period People's View of Earthquakes," in which he states that the common people in 1855 thought that movements of a giant catfish under the ground caused earthquakes.[111] In his study of the 1891 Nōbi earthquake, Gregory Clancey claims, "Before the Meiji Restoration . . . earthquakes were considered the consequence of movements of a giant catfish."[112] Pioneer seismologist Musha Kinkichi also wrote that "people of old" in Japan regarded a catfish as the cause of earthquakes, although he explains that the idea originated at some point after the Kamakura period. Calling the idea that catfish cause earthquakes "ridiculous," Musha then cites a tale from the 1856 *Ansei Chronicle* (*Ansei kenmonshi*) as a segue to exploring the possibility that catfish are able to predict earthquakes.[113] Clearly, it is common for scholars in a variety of disciplines to believe that in the pre-modern past, a significant subset of Japan's residents thought that catfish caused earthquakes. Some even claim that this notion is of ancient vintage.

A closer look at earthquake-related literature during the Tokugawa period, however, calls this claim into question. First, although some of the literature on earthquakes discussed catfish in the context of earth-quake folklore, no academic work ever regarded catfish as a cause of earth-quakes. There was no academic "catfish theory" of earthquakes during the Tokugawa period, nor was there a single, coherent folk theory involving catfish. The popular press consistently attributed earthquakes to imbal-ances in yin and yang, usually providing a simplified version of academic explanations. A typical example is a broadside print describing the 1853 Odawara earthquake as a clash of yin and yang, causing "thunder and rain in the skies and an earthquake on the ground."[114] There is no doubt that by the nineteenth century, catfish had become a widely known *symbol* of earthquakes, but I have found no direct evidence that anybody regarded this symbol as the actual causal mechanism of earthquakes.

The catfish print entitled *Compassion of the World-Rectifying Catfish* (*Yonaoshi namazu no nasake*) features a catfish defending his species, saying rhetorically, "Even if millions of fish were to advance, how could we move this great earth even one inch?" Moreover, "Earthquakes are yin and yang energy," says the catfish, so people should not think badly of us.[115] The print is not seriously trying to explain earthquakes but is instead relying on a jux-taposition of the prevailing commonsense view that imbalances in yin and yang cause earthquakes with the catfish symbol to make a mildly humorous point. In general, the 1855 catfish prints relied on the metaphor of the move-ment of a giant catfishlike creature pinned down under the Kashima Shrine as the cause of earthquakes. This flexible metaphor enabled the prints to make statements about a wide variety of social phenomena while ostensi-bly discussing the earthquake. There is circumstantial evidence that even in 1855, the story of the giant catfish pinned down by the Kashima deity push-ing on the Foundation Stone was not universally known among the resi-dents of Edo. We cannot say with certainty what was in people's minds, but Tomisawa Tatsuzō points out that many early catfish prints featured detailed explanations of the mechanism by which Kashima suppresses the giant cat-fish with the Foundation Stone as well as other relevant items of earthquake lore.[116] In other words, at least some consumers of the prints, perhaps many of them, needed an explanation to grasp fully the relevant symbolism.

When catfish prints discussed the actual causes of earthquakes, they relied on the workings of yin and yang, sometimes enhanced by Buddhist or other ideas. One print entitled *Earthquakes Explained* (*Jishin no ben*) features the image of a dragonlike creature encircling Japan—a common variation of the catfish metaphor. The print summarizes earthquake-related phenomena at considerable length. After explanations based on yin-yang, the clash of fire and water, and the geological characteristics of the earth, the discussion turns to earthquake lore:

> According to folklore (*zokusetsu*), there is a giant catfish under the ground. The movement of its head and tail causes the earth to shake. Foregoing a detailed discussion of origins, there was a creature called an earthquake insect (*jishin no mushi*) on the face of an almanac from 1198. Written upon its figure are the sixty-six provinces of Japan. This explanation of earthquakes comes from six or seven hundred years ago, and in Buddhist sutras, it is a dragon (*ryū*). As *Thoughts on Earthquakes* [*Jishinkō*, 1830] states, this image is of ancient vintage. In examining various books from the time, the country did not always rest atop this creature, nor did the creature always assume the guise of a catfish. There were numerous variations on the dragon theme. We have borrowed this image for use in this print.[117]

Again, we see no suggestion that anyone regarded the folk explanation literally, even in a print produced for mass consumption. Moreover, the anonymous print author displays a sophisticated grasp of the mutability and complex history of the hybrid dragon-catfish-insect creature associated with earthquakes.

The main evidence that some Japanese may have believed a catfish or dragon under the ground caused earthquakes is that a small number of authors criticized the idea with sufficient vigor to suggest that somebody may actually have taken it seriously. For example, Kojima Fukyū wrote as follows in *Treasury of Lamp-Lighting Dialogues* (*Heishoku wakumonchin*, 1710) under the heading "Earthquake Explanation":

> *Question:* According to the explanations of women and children, the Kashima deity regrets earthquakes, and he uses the Foundation Stone to pin down a catfish. Have you heard of such a thing?

Answer: Occasionally yin and yang energy creates violent forces that can collapse houses and injure people. . . . As for the Foundation Stone at Kashima, in the talk of women and children there is a giant catfish under the earth upon whose back the entire territory of Japan rests. The movement of its head or tail causes violent shaking of the earth. For this reason, the Kashima deity pushes down on the Foundation Stone. Giving this matter some thought, how could this catfish support only Japan, not China, and yet China, too, is subject to earthquakes? It is laughable.[118]

In 1792, Takai Saiga was similarly dismissive in a work aimed at general audiences. In *Primer on Heaven and Earth* (*Kunmō tenchiben*), he wrote sarcastically, "We can say that explanations such as the catfish are the result of people who think that no countries exist outside Japan and who have no sense of the vastness of heaven and earth. China, India, and Europe all have large and small catfish. Does a fish surround these countries? If so, then every one of the myriad countries should have its own Foundation Stone."[119] With the exception of relatively rare passages like these, few writers devoted space to rebutting the idea that catfish cause earthquakes. The topic typically came up in the context of earthquake lore, and serious intellectual attention to it sought to explain its folkloric origins or the reasoning behind the metaphor.

It is possible, therefore, that some Japanese in the eighteenth and nineteenth centuries thought that catfish or other aquatic creatures caused earthquakes or were connected with earthquakes in some important way. Folk beliefs can be notoriously difficult to document. Nevertheless, all evidence leads to a conclusion that the majority of early modern Japanese understood the metaphoric nature of the earthquake catfish.

The remainder of this book consists of five chapters. Chapter 2, "Why the Earth Shakes," examines early modern understandings of the causes of earthquakes from the standpoint of intellectual history and the history of science. Chapter 3, "Japan according to Earthquakes," is arranged topically. Drawing on material from more than a dozen major earthquakes, I analyze their social effects. Next, "The Ansei Edo Earthquake" examines that event in detail, including patterns of destruction, patterns of relief, and some of its cultural products. After examining the earthquake itself, chapter 5, "Meanings," analyzes its social, political, and economic significance, especially in light of the collapse of the bakufu twelve years later. Chapter 6, "Into the Twenty-First Century," examines the iconic function of the Ansei Edo earthquake in the modern era and the influence of early modern earthquake lore on the modern science of seismology, conceptions of Japan, and contemporary life in a country where the destructive potential of earthquakes and tsunamis has been dramatically demonstrated in the recent past.

CHAPTER 2

Why the Earth Shakes

Late in 1662, Asai Ryōi wrote *Foundation Stone* (*Kaname'ishi*), a work of fiction in the guise of a journalistic account of the Kanbun earthquake. In an effort to conclude on an uplifting note, Ryōi explained that according to a poem by the Song Chinese intellectual Su Dongpo, the month in which an earthquake occurs determines fortunate or unfortunate events in the near future:

> The prosperity of the people declines, the spring is "fire" [= the fire or hearth deity]; severe droughts occur
>
> The second, fifth, and eighth months are "dragon" [= the dragon deity]; people of high status and low all perish
>
> The sixth, ninth, and first months are "gold" [= the golden-winged bird]; the harvest of rice and other grains is abundant
>
> The seventh and twelfth months are "emperor" [= celestial king]; wars and rebellions will occur.

Therefore, this year's earthquake signals an abundant harvest of the five grains and prosperity.[1] Even the great sage kings of ancient times, whether in China or our country, could not avoid occasional disasters owing to the workings in yin, yang, and the five phases. It is only natural that such occurrences could happen today and should not be cause for alarm. On the contrary, at present, with the seas in all directions calm and society well governed, events like the present earthquake can be regarded as reports from the deities revealing the future, not as unfortunate happenings.

In popular lore, there is the notion that the world is supported by a dragon king, whose anger is the cause of earthquakes. The Kashima deity suppresses this dragon king, whose head and tail are twisted and overlap at one point. Because the Foundation Stone is located above this point, no matter how violently the dragon king shakes, it will not destroy human society. An old saying goes: "The Foundation Stone will not be thrown off no matter how great the shaking, as long as the Kashima deity is present." Therefore, I name this record of the earthquake "Foundation Stone."[2]

Here yin, yang, and the five phases mix with numerology, the will of deities, popular monsters, and the prestige of a famous Chinese poet to explain earthquakes and reassure readers that the present disaster is a normal yet rare occurrence in the larger picture of a peaceful, well-governed era. Variations on the old saying Ryōi cites about the Foundation Stone and social stability appeared in the literature after every major early modern earthquake. Such invocations of stable foundations served as a reassuring formula.

Notice that although Ryōi cites the five phases of yin and yang as the physical mechanism causing the earth to shake, the intentions and actions of *kami* (deities) also play a role, albeit one Ryōi does not articulate clearly. Although Ryōi obviously intended to put a positive spin on events, his claim of *kami* causing a terrifying, deadly event as a way of reporting that good fortune lies ahead hearkens back to a much older notion of *kami* as capricious, violent, and beyond human control.[3] Of course, Ryōi's purpose in writing was to sell books, not to achieve rigorous academic consistency. One importance of *Foundation Stone* for a study such as this one is that it displays many of the possible explanations of earthquakes that had developed by the middle of the seventeenth century, a time when older and newer theories mixed and a clear consensus had yet to emerge.

As Ryōi's writing indicates, there were many dimensions to the perceived causes of earthquakes. This chapter focuses on scientific thought, broadly defined, with some attention to religious matters, especially cosmology. I argue that with respect to earthquakes and the broader world of ideas, there was no sudden paradigm shift at either end of the early modern era. Periodization based on government institutions does not necessarily correspond with other realms of human endeavor. Buddhist cosmology and other ideas common in the medieval era carried over into the seventeenth century, where they mixed with Chinese-derived cosmology and certain European ideas. Older concepts never entirely faded, but by the eighteenth century a consensus emerged that imbalances in yin and yang energy were the proximate physical cause of earthquakes. Though the details varied from writer to writer, the most common basic scenario was that yang energy accumulated under the ground and sought to rise. If soil and other conditions impeded its natural propensity to rise, the pressure from this yang energy built up to the point that it exploded. The usual mechanisms posited for yang energy entering the ground, ordinarily the realm of yin, were wind and heat from the sun entering natural openings in the earth.[4]

I further argue that the Ansei Edo earthquake called this yin-yang view into question, but no obviously better theory was available as a radical replacement. Alternative theories such as the buildup of electrical charges under the earth or the ingredients of gunpowder mixing under the earth closely resembled the mechanism of yin-yang in key respects. The lack of a clearly superior explanation for earthquakes at the start of the Meiji era is one reason that older ideas about earthquakes influenced the modern development of seismology. To contextualize the situation in the early Tokugawa period, I briefly discuss the medieval notions of natural disasters. I take up the situation in the Meiji era and beyond in the final chapter.

Early and Medieval Theories of Natural Hazards and Disasters

The general term for the Chinese idea that natural calamities were heavenly warnings or punishments for a lack a virtue on the part of the ruler is *zaiyi sixiang* (Japanese: *saii shisō*). The *zai* of *zaiyi* included events such as droughts, floods, famine, insect infestations, fires, and rebellions, and *yi* indicated portents such as eclipses, earthquakes, unseasonable weather, and abnormal plants and animals. This basic concept was present in Japan by the time of the formation of the Ritsuryō state in the eighth century, and it coexisted with an alternative understanding of natural and social disasters as curses (*tatari*) from the *kami*. Because early conceptions of the *kami* usually regarded them as inscrutable and arbitrary in their violence, this alternative understanding conflicted with the view that natural disasters were signs the sovereign or state was deficient in virtue.

Early Japanese emperors typically took responsibility when natural hazards caused major loss of life or destruction, but formal declarations typically deemphasized any lack of virtue on the part of the sovereign. Instead, the focus of imperial responsibility was enacting corrective measures such as intoning sutras at major temples. The typical arsenal of corrective actions included donations to major shrines and temples, mobilizing temple and shrine priests, declaring a general amnesty, and distributing rice and salt to the poor. Other measures might include rebuilding destroyed dwellings, ensuring proper burial of the dead, exempting those above age sixty from labor service, imperial edicts exhorting officials to be mindful of their duties, and changing the era name.[5] Some of these elements remained part of official responses to calamities even during the nineteenth century. In the wake of the Ansei Edo earthquake, for example, the bakufu mobilized the religious power of shrines and temples and provided material relief to retainers and townspeople.

During medieval times, Buddhist cosmology tended to condition thinking about natural hazards and disasters. The basic idea was to link good or evil behavior in the realm of humans with the balance of power between good and evil deities. In this view, the conflict between good and evil deities controlled natural phenomena on earth.[6] Japan's *kami* became avatars of buddhas and bodhisattvas, and, as such, the *kami* became more rational in the sense of behaving predictably according to moral standards. Insofar as human behavior could influence the *kami* for better or worse, it could increase or reduce the severity of natural hazards. In *Treatise on Securing Peace for the Country* (*Risshō ankokuron*, 1260), for example, Buddhist thinker Nichiren explained that when people turn their back on the True Law and embrace the Wicked Law, a variety of problems plague a country. These problems include abnormalities of the sun and moon, abnormalities of the stars, fires, floods, windstorms, droughts, external military invasions, famine, warfare, and epidemic disease. In one part of the essay, Nichiren explains that when the principles of Buddhism become obscured and lost, "at that time, loud noises will sound in the air and the earth will shake; everything in the world will begin to move as though it were a waterwheel. City wells will split and tumble, and all houses and dwellings will collapse. Roots, branches, leaves, petals, and fruits will lose their medicinal properties. . . . The crops will all wither and die, all living creatures will perish, and even the grass will cease to grow any more. Dust will rain down until all is darkness and the sun and the moon no longer shed their light."[7]

In Nichiren's *Treatise* and many other Buddhist writings, earthquakes are part of a larger package of possible calamities including the complete destruction of human society.

The most common Buddhist texts deployed in attempts to prevent or minimize the effects of malevolent cosmic forces were the *Heart Sutra*

(*Daihannya-kyō*) and the *Benevolent King Sutra* (*Ninnō-kyō*). In 1231, for example, the Kamakura bakufu ordered thirty priests to intone the *Heart Sutra* at the Tsuruoka Hachiman Shrine at the height of the Kangi Famine.[8] Another common text for guarding against natural disasters was the *Golden Light Sutra* (*Konkōmyō-kyō*).[9]

Early Modern Academic Theories of Earthquakes

Between 599 and approximately 1990, over forty-five thousand instances of felt earthquakes appear in extant records. Approximately 1.1 percent of these records are from the period of 599 until 1600, and 89 percent of them are from the Tokugawa period.[10] This sharp increase in the documentary coverage of earthquakes corresponds with the emergence in the seventeenth century of specific theories about their causes. During medieval times, earthquakes were but one variety of "earthly anomalies" (*chii*). During the Tokugawa period, earthquakes became an object of academic study in their own right. Although there was considerable overlap and cross-influence, academic theories of earthquakes tended to fall into three categories based on the origins of their theoretical frameworks: Buddhist, Western, and Chinese. Theories based on a Chinese-derived yin-yang framework became most influential as time went on.

Basic Buddhist Theory

During the seventeenth century, Buddhist cosmology often supplied material to explain earthquakes. The anonymous 1662 *Supreme Ultimate Earthquake Record* (*Taikyoku jishinki*), another product of the Kanbun earthquake, was particularly influential. Although it explains earthquakes from a Buddhist cosmological perspective, the work also contains extensive quotations from Confucian and Daoist writings. Buddhist explanations of earthquakes typically start with the Five Seed Elements of earth, water, fire, wind, and space. As for the structure of the earth, "above space is wind. The agitation of wind causes it to blow upward, likewise agitating fire. When fire becomes well agitated, water is no longer calm, and it functions as a substance [*ki*] that can long support things floating on it. Therefore, it is easy for the earth to float atop water. The power of the fire and wind below causes the water to surge upward, and the earth serves as a lid to hold the water down."[11] The element space is found in the heavens, and the deepest layer of the earth is wind. This wind feeds fire, the next layer, which boils the water above it, which supports the earth. The above passage describes the normal state of affairs, with upwelling forces and downward suppressing forces in equilibrium, much like yin and yang balancing each other.

What preserves this balance? According to *Supreme Ultimate Earthquake Record,* it is "the ki of the two *kami,*" with ki indicating matter or material force (Chinese: *qi*) in the context of Chinese cosmology and the *kami* as willful cosmic forces. Ominously, "When the people offend against the will of the *kami,* that offense penetrates heaven and earth." The anger of the deities whips up a divine wind that can manifest itself as earthquakes, typhoons, floods, severe waves, or poor agricultural conditions—among other problems.[12] Notice here the key role of human behavior in influencing the cosmic forces. In this respect, little had changed from medieval conceptions of natural hazards.

Supreme Ultimate Earthquake Record explains the basic mechanism of earthquakes in several variations, striking what we might call a scientific tone in some places and focusing on moral qualities in others. For example, "Earthquakes are the result of abnormal transformations within the earth, water, fire, and wind." After proposing the metaphor of water boiling in a pot, the text explains that the mechanics of earthquakes are the same everywhere under heaven.[13] Other metaphors come into play to make the same basic points, reiterating that earthquakes, typhoons, and floods share the same root cause, namely "abnormal transformations of ki."[14] In what became a common feature in essays on earthquakes during the Tokugawa period, *Supreme Ultimate Earthquake Record* provides a list of fifteen major earthquakes starting with one in 416, the earliest recorded earthquake in the Japanese islands, and ending with the Kanbun earthquake of 1662.[15] Subsequent paragraphs describe earthquake lore, reiterate the point about earthquakes being caused by transformations of ki, and point out that space or emptiness is the root of the way.[16]

The final paragraph in *Supreme Ultimate Earthquake Record* includes the rhetorical question, "If we reject the seeds of ethics and promptly forget the way of the *kami,* will not the state perish?" Subsequent sentences shift the focus from the negative consequences in the question to the benefits of correct behavior and mindset. The dominant metaphor is a mirror, whose essence remains unchanged and that reflects reality without distortion or bias. The final image is of the sun and moon shining brightly, in harmony with people whose pure minds reflect divine virtue. In such a state, "What fear need we have of earthquake anomalies?"[17]

This ending is similar Ryōi's attempt to reassure readers tha⌐ looks bright. In another similarity between the two contemporaneous ⌐ Ryōi provides a brief description of over twenty past earthquakes starting in 416, including some that occurred in China. Immediately after this description, there is a concise explanation of the Buddhist view of earthquakes, which differs slightly from *Supreme Ultimate Earthquake Record*. Wind is the deepest layer in Ryōi's summary, but instead of nurturing fire, the wind agitates a layer of water. The gold layer is located above the layer of water, and earth forms the outermost layer. Vigorous movement of the wind can agitate all of the other layers, thus causing an earthquake. Ryōi's explanation is entirely mechanical, with no mention of moral issues or human behavior.[18]

European Ideas about Earthquakes

Greek thinker Anaximenes of Miletus (586–528 BCE) saw the earth as full of holes, like a sponge. Rainwater filled these holes and accumulated. When obstructions blocked the rising waters, the accumulated pressure could cause earthquakes. Moreover, when this water dried up, it caused the earth to crack, which caused mountains to collapse, making the earth shake. Democritus (ca. 460–ca. 370 BCE) thought that groundwater displacement was the basic cause of earthquakes. In his view, the earth was originally full of water and excessive rains might cause this water to overflow, thus triggering earthquakes. When parts of the earth dried out, large quantities of water rushing into the dry area could also cause earthquakes. Aristotle (384–322 BCE) was doubtful of these explanations. For one thing, even constantly dry regions experience earthquakes, and if Anaximenes was correct, the earth's surface would gradually become more level and the frequency of earthquakes would diminish.[19]

Aristotle's view of earthquakes came to Japan via Christian missionaries in the late Muromachi period and remained influential during the Tokugawa period. For Aristotle, wind was the main cause of earthquakes, and he saw the world as consisting of five elements: fire, earth, air, water, and ether. This obvious resemblance to theories of earthquakes based on Buddhist cosmology, along with imprecise translation, resulted in theories of earthquakes based on Aristotle often appearing to be of Buddhist origin.

The first systematic explanation of earthquakes in Japan was *Alpha and Omega Explained (Kenkon bensetsu)*, first written in roman letters in 1650 and rewritten in Japanese script in 1659. "*Kenkon*" is the Japanese pronunciation of the Chinese *qian-kun,* the pure yang and pure yin tri-grams in the *Classic of Changes* (*Yijing*). The original author of *Alpha and Omega Explained* was Sawano Chūan, formerly Christovao Ferreira, who lived in Japan for forty years. His work was based on a European astronomy book that a Japanese official obtained from Portuguese priests undergoing interrogation in 1643. True to its title, the work explains a wide variety of natural phenomena. One chapter deals with earthquakes, and the explanation is basically Aristotelian: "That which we call an earthquake is the result of bursts of wind entering the large holes in the earth and becoming trapped there. Because this collected wind is located under subterranean water, when it tries to escape from the earth it can find no suitable path. The force of the wind trying to escape causes the earth to shake."[20] There is an obvious resemblance to the action of wind and water previously described in Buddhist theories.

The work goes on to specify additional mechanisms. One possibility is that when wind does escape through the earth's large holes, the air below has the qualities of being warm and damp, whereas the earth has the opposite qualities of being cold and dry. The conflicting qualities can cause the earth to shake. In cases when the opposite qualities blend harmoniously, the escaping yang-air (called *eisarasan*) becomes hot and dry. When wind blows anew, this new wind and the escaping hot and dry air are in conflict, and the new wind pushes the escaping air back, causing the earth to shake. Another possible cause of earthquakes is a large quantity of this accumulated hot and dry air. Because it naturally seeks to rise, this air can cause the earth to shake. Islands and areas along rivers are especially prone to earthquakes owing to these forces.[21]

Many Tokugawa-era scholars were aware of both classical and more recent European theories of the earth and planets. The prolific philosopher and astronomer Miura Baien (1723–1789), for example, made reference to the ideas of Hipparchos (second century BCE), Ptolemaios (Claudius Ptolemy, second century), Nicolas Copernicus (1473–1543), and Tycho Brahe (1546–1601).[22]

Theories Based on Chinese Concepts

Even discussions of earthquakes based on Buddhist cosmology or Aristotelian views usually employed a Chinese-derived metaphysical vocabulary to explain the workings of heaven and earth. The Chinese text with the greatest influence on theories of earthquakes in Tokugawa Japan was *Astronomy Questions Answered* (Japanese: *Tenkei wakumon;* Chinese: *Tianjing huowen,* 1597). Although mainly about astronomy, the second volume of this work contains a discussion of earthquakes. *Astronomy Questions Answered* became widely read after 1730, when Nishikawa Seikyū added Japanese reading marks to the classical Chinese text. This work circulated among scholars early in the Tokugawa period, even though technically it was a prohibited book until Tokugawa Yoshimune's reign as shogun (1716–1745). Based on their content, it is clear that several Japanese discussions of earthquakes prior to 1730 relied on the explanation in *Astronomy Questions Answered.*[23]

Reminiscent of Aristotle's view, *Astronomy Questions Answered* explains that there are holes or spaces in the earth, similar to the appearance of a bees' nest or the cap of a mushroom. Therein lies concealed the material substance of fire and of water, the violent collision of which produces earthquakes when there is no outlet. The same process produces thunder. Places with sandy or muddy soil such as the North Pole or the equator rarely experience earthquakes. On the other hand, warm, rocky areas are the ideal environment for earthquakes. When earthquakes occur under buildings, it is like an explosion of gunpowder, causing much damage.[24] The force of earthquakes is capable of causing mountains to shift and the surface of the earth to tear. These forces can create new mountains or islands and reverse the flow of rivers. When the violent material force dissipates as heat, the shaking stops.[25]

One classical Chinese book many Japanese authors mentioned is the *Classic of History* (*Shijing*), which is the source of the basic idea that rising yang energy causes earthquakes: "When the yang energy is trapped and cannot emerge, or the yin energy represses it and keeps it from rising, then there is an earthquake."[26] The *Classic of History* did not elaborate on details such as how the yang energy became trapped, so later Chinese and Japanese books fleshed out this basic idea. Terashima Ryōan began the discussion of earthquakes in his ca. 1712 encyclopedic work, *Illustrated Compendium of Chinese and Japanese Knowledge* (*Wakan sansai zue*), by stating "When yang exists below yin and is suppressed by yin, if there is no way for yang to rise upward, an earthquake occurs." Next, we find a paraphrase of *Astronomy Questions Answered.* According to "astronomy books," the earth is like a bees' nest or the cap of a mushroom in that interconnected holes or conduits permeate it all around. Water force and fire force reside in these conduits. When the energy from the accumulation of these forces cannot erupt to the surface, it is like a cramp developing in a person or thunder in the skies.[27]

Ryōan explains that extremes of climate result in fewer earthquakes. Far northern lands are so cold that they do not generate heat, and lands below the equator readily dissipate heat because the sun is so strong. In rocky areas at temperate latitudes, by contrast, there are spaces beneath the surface into which hot energy blows and is trapped by cold energy. When this situation surpasses a threshold, an earthquake occurs. Large earthquakes do not cause uniform shaking over a wide area but instead cause localized upwelling of energy. Sometimes the underground energy turns into fire owing to intense heat. After it departs to the surface, the shaking stops.[28]

After conveying the *Astronomy Questions Answered* explanation of the cause of earthquakes, Ryōan elaborates further. Normally, yin and yang are in balance within the earth, but a blockage of yang leading to its long-term accumulation can cause the earth to expand and water to contract like a rice cake expanding when cooked. Wells can go dry or the water in them becomes warm as a result. Possibly referring to different types of seismic waves, Ryōan points out that when earthquakes strike, the initial shaking is especially severe but the second wave is weaker. He then explains that "black waves [*kuronami*] . . . commonly known as tsunami," as high as a mountain can occur when an earthquake stirs up ocean water by the expansion of land. After a large earthquake, lesser shaking can occur for months afterward because the trapped hot energy has not fully escaped. Ryōan explained volcanic eruptions as another manifestation of the same basic situation of hot, yang energy trapped underground. He points out instances in 863/864 and 1707 of Mt. Fuji erupting soon after an earthquake and one case in 800 of an eruption not preceded by an earthquake. In this case, Ryōan speculates that the eruption released all of the underground energy.[29]

Ryōan continues the discussion of earthquakes by pointing out that although all countries exist under the same sky, earthquakes are localized phenomena. An earthquake in China will not be felt in Japan, nor will one in Osaka be felt in Edo, and the severity of shaking varies even within the same village as a function of whether the soil is firm or soft.[30] This last point was an advance in knowledge, and by the end of the Tokugawa period we find variations of this topic in a range of literature. For example, after the 1855 Ansei Edo earthquake, Jōtō Sanjin explained in *Account of Broken Windows* (*Mado yabure no ki*), "In the current earthquake, high ground shook and low ground shook severely. The situation is that Azabu, Yotsuya, Hongō, Komagome and nearby high ground shook, and the castle grounds, Ogawamachi, Koishigawa, Shitaya, Asakusa, Honjo, Fukagawa, and nearby areas shook to a greater degree. This is a natural principle [*shizen no kotowari*]."[31] Even more to the point was Mishima Masayuki in *Observations after the Earthquake (Nai no nochimigusa)*. Walking through Edo in the days following the Ansei Edo earthquake and recording the damage in each area, combined with his knowledge of the city's past, Masayuki pointed out that areas such as "Daimyo Lane" were built on land that long ago had been reclaimed from the ocean. "Such places," he deduced, "naturally [*onozukara*] shake more severely in an earthquake."[32] Modern textbooks point out that ground motion is most severe in the case of unconsolidated fill, owing to an increase in wave amplitude, a vertical shift in the path of seismic waves, and because waves are trapped by basement rock.[33]

The final point in Ryōan's discussion is instructions for a simulation that "might seem to be child's play, but very closely simulates the basic logic of earthquakes." In a bucket, pile up some rough sand with a spout at the bottom and add water. Several people in rotation blow air through the spout. Initially the air does not emerge because yin has trapped yang. Eventually the bucket will shake. When the air has escaped, the shaking will stop.[34] Explicitly citing *Illustrated Compendium of Chinese and Japanese Knowledge*, the popular 1856 work *Ansei Record (Ansei kenmonroku)* included this same simulation.[35]

The observations Ryōan passed along concerning seismic waves and varying degrees of ground motion in different localities and soil bases were potentially valuable contributions to an emerging science of earthquakes. The precision of the empirical observations of some Tokugawa-period writers was remarkable. For example, *Kokuryōki,* an account of conditions in the Tosa domain in Shikoku following the powerful 1707 Hōei earthquake, contains the following entry: "After the great earthquake, the contour of the earth [*chikei,* topography] rose at the harbors of Tsuro and Murozu in Aki-gun. Last year, large ships full of cargo could enter the port freely, but after the disaster, large ships full of cargo were unable to enter. Because this port was cut from rock, the bottom consists of rock, so it is not a case of mud filling in the bottom. [This change] is proof that the contour of the earth has risen."[36] Uplift and subsidence were common occurrences in ocean trough (Tōkai and Nankai) earthquakes such as Hōei.

Another influential work that relied heavily on Chinese theories was Nishikawa Joken's 1715 analysis of heavenly and earthly phenomena, *Strange Phenomena Explained (Kaii bendan).* The work consists of four "heaven" volumes and four "earth" volumes, with earthquakes discussed as the first topic in the first earth volume. The initial paragraphs cite Chinese sources from ancient times through the Ming dynasty, including the *Classic of History* explanation of accumulated yang trapped by yin as the cause of earthquakes. Some of the passages provide analogies between natural phenomena and human phenomena. For example, *Strange Phenomena Explained* likens the shaking of the earth to disordered relations between rulers (yang) and ministers (yin) and the resulting upheavals. Several passages also mention specific large earthquakes in China and the extent of casualties they caused. Following these Chinese accounts is a brief record of early Japanese earthquakes between 845 and 870.[37]

Joken's own analysis (the *bendan*) then begins with the observation that only accounts of major earthquakes are found in the historical record, and these events are rare in both China and Japan. Taking up the common Chinese idea that astrology can predict earthquakes, Joken argues that destructive earthquakes may occur so suddenly that predicting or divining them is impossible. In the classical process of divination, earthquakes occurred by yin filling in a loss or absence of yang, and therefore the vigorous activity of ministers of state was one possible indication of an impending earthquake. If, however, we follow the explanation in the *Yellow Emperor's Classic of Medicine,* the action of wind causes earthquakes. Wind is yang energy that finds its way into the earth and seeks to flow and expand. Earthquakes happen when yin energy suppresses this wind. After discussing an analogous process as a cause of illness within the body, Joken declares that the wind theory and the theory of yin filling in for absent yang are substantially the same. Next, he mentions the Buddhist idea of the earth sitting atop a layer of water, which wind can agitate to the point of causing the earth to shake. He concludes that the Buddhist theory is largely the same as the causal mechanism mentioned in *Yellow Emperor's Classic of Medicine.*[38]

Joken's next topic is the popular notion of an "earthquake fish" (*jishin no uo*). By 1715, the specific association of catfish (*namazu*) and earthquakes had developed, especially in the area around Lake Biwa, but catfish had not yet become a widespread symbol of earthquakes. By the early eighteenth century, the general idea or metaphor of some sort of giant dragon, serpent, *mushi* (worm/insect), fish, or other creature moving about under the earth was common in Japanese folklore.[39] Joken points out that fish are yang creatures living within a yin environment. Therefore, their association with earthquakes was probably because fish could serve as a metaphor for the Buddhist notion of wind (yang) agitating water (yin). "In any event,"

Joken concludes, "this notion is far from correct principles." After taking up the fish topic, Joken concludes the section on earthquakes by mentioning a country named "Berū" (Peru), located to the southeast of Japan, which is subject to many severe earthquakes. Moreover, there are reports of many other countries in the world in which earthquakes are frequent.[40] He may have mentioned the occurrence of earthquakes in other countries immediately after the fish discussion because most folklore about an earthquake-causing creature imagined it to be located under Japan.

The next section is entitled "Fissures in the Earth and Fissures in Mountains," two conditions caused by large earthquakes. Here too, Joken begins with specific examples from classical Chinese sources followed by Japanese sources.[41] His own analysis states that these phenomena follow the same logic as earthquakes, namely that yang energy in the earth provides the force to break apart earth and rocks owing to the shaking. Joken then points out the phenomenon of liquefaction, whereby mud sometimes issues from cracks in the earth. His final point is that "something resembling ash" sometimes also issues from these cracks, depending on the situation of the earth at that location.[42] Joken's discussion of earthquake-related phenomena continues in sections entitled "Subsidence of the Earth and Mountains," "Collapses of Mountains," "The Emergence of Flat Land and Islands," "The Movement of Mountains, Mounds, and Islands," and "The Groaning of Mountains." These sections follow the same approach as the ones examined thus far: examples of specific cases from classical texts, an analysis based on the basic idea of yang in the earth seeking to escape upward, and mention of the phenomenon in places other than Japan and China.[43]

Joken published a similar eight-volume work in 1712, *Discussions of the Heavens* (*Tenmon giron;* also known by the title *Ryōgi shūsetsu*). In this earlier work, his basic theory of earthquakes is the same as *Strange Phenomena Explained.* One difference is that the earlier work paraphrases *Astronomy Questions Answered* in several places.[44] Joken's major accomplishment in these works was to synthesize Buddhist, Western, and Chinese ideas about earthquakes and apply a basic causal mechanism to explain a wide variety of earthquake-related phenomena. The general idea that imbalances in yin and yang were the physical mechanism causing the earth to shake became the ordinary, commonsense understanding of earthquake mechanics during the eighteenth century.

In the broader perspective, the idea that yang accumulates in the earth and moves upward toward the sun functioned as more than an explanation for earthquakes. This idea was a foundational concept in many general theories of the earth. Miura Baien, for example, explained in the middle of the eighteenth century that the ground is a concentration of yin energy from which yang emerges. Under certain circumstances, this upwelling of yang breaks up the earth's surface, thus creating areas for water to accumulate.[45] Indeed, a complex interplay between yin, yang, and its five agents explained the composition and movements of the earth, moon, and sun for Baien and many other Tokugawa intellectuals. Let us now move into the nineteenth century to examine several influential texts.

The 1830 Kyoto earthquake caused several hundred deaths and several thousand injuries, and it was followed by an unusually large number of aftershocks. It was the first major earthquake to shake the imperial capital in living memory. Major earthquakes had sometimes shaken Kyoto much earlier, and Asai Ryōi's *Foundation Stone* describes in detail the destruction to the city in 1662 following the stronger and more deadly Kanbun earthquake. Although the 1662 event was more powerful and destructive, the 1830 earthquake seems to have generated more social anxiety. One reason was the timing. It was an *okage* year, featuring popular religious pilgrimages with millenarian overtones. Whereas the Kanbun earthquake shook a wide area including Kyoto, the 1830 earthquake shook mainly the city itself. Most important, the popular press was much more prominent than in 1662. Competition for sales of prints and ephemera resulted in exaggerated reports of death and destruction. This coincidence of circumstances garnered much attention throughout Japan for the 1830 Kyoto earthquake. Among its many products were two influential books.

The more important book was *Thoughts on Earthquakes* (*Jishinkō*). Kojima Tōzan wrote it with the explicit purpose of calming fears of a breakdown in social order by explaining earthquakes in a rational manner. Although "Tōzan sensei" appears prominently on the cover as the author, the book appeared after his death. Tōzan indeed wrote the first half, but an anonymous student, known only by his pen name Tōrōan-shujin, wrote the more interesting second half.[46]

Thoughts on Earthquakes begins with a basic description of the earthquake that had just shaken Kyoto. It points out that the shaking damaged or destroyed many storehouses and earth embankments and some houses, and injured many people. Tōzan explains that while events like this had happened long ago, nothing on such a scale has taken place in recent years. "People are shocked and terrified." To reassure readers, Tōzan quotes a proverb: "Earthquakes are severe at the beginning, typhoons are at their worst in the middle, and thunder is dramatic at the end." The point, of course, is that the worst is now behind us. Next, "for the purpose of calming people," he noted that "since antiquity, historical records in Japan have recorded the occurrence of earthquakes." His first example is from 887, and the context of subsequent examples indicates that they are all earthquakes that shook Kyoto. Unlike the worldwide scope of authors such as Nishikawa Joken, Tōzan's approach was to normalize the 1830 earthquake by putting it into the context of a long line of such events affecting the imperial capital. Another calming function of these examples was to point out that aftershocks, indicated by counts of the number of times shaking occurred, are a normal part of earthquakes. At the end of the section, Tōzan explains that in each series of shaking he described, the first instance was always the most severe and the intensity decreased afterward.[47]

The next section is a basic explanation of earthquakes, and it contains little that was new at the time. The first statement is that earthquakes occur when yang energy within the earth seeks to rise but is trapped by yin energy. The next paragraph cites *Astronomy Questions Answered* and explains that there are holes in the earth like a bees' nest or the cap of a mushroom and that the contact of fire and water creates a force that seeks to rise and causes the earth to shake if blocked. Next is the familiar geographical information whereby the cold extreme north and hot equatorial areas as well as areas of sandy or muddy soil experience few earthquakes. By contrast, warm areas with rocky soil suffer the highest frequency of earthquakes. The gunpowder analogy of *Astronomy Questions Answered* got a slight upgrade in Tōzan's account, becoming the blast of an old-style cannon.[48]

The explanation of the causal mechanism of earthquakes is brief, and a longer section on the signs of earthquakes follows. This point segues into folklore concerning fish and earthquakes. *Na* refers to a fish (*uo*), and *e* refers to shaking (*yuri*). *Nae,* the old name for earthquakes, therefore, may have derived from the metaphor of a captured fish vigorously shaking. Tōzan's language clearly indicates the metaphoric nature of the fish-earthquake connection, which he describes as "truly the explanation of a small child." He then turns to the topic of an illustrated almanac from 1198 featuring an "earthquake insect" looking like a giant centipede surrounding all the provinces of Japan (fig. 2). This item is now widely regarded as a seventeenth century forgery, although Tōzan would not have known.[49] He points out that in Buddhist explanations, dragons were linked with

FIGURE 2 The "earthquake insect/caterpillar" (*jishin mushi*) as it appears in *Thoughts on Earthquakes* (*Jishinkō*, 1830). An icon of Mt. Fuji appears in the center of the provinces it encircles.

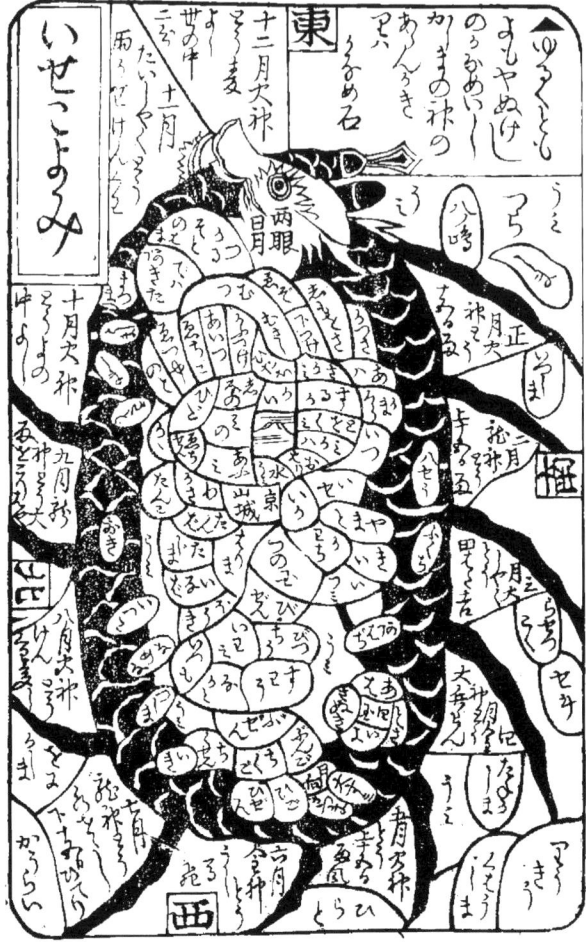

earthquakes much like fish were in Japan. In any case, he says, there is a variety of old lore about earthquakes. Tōzan's section of the book ends by stressing that in the "current peaceful age," this earthquake is not a sign of anything greater than itself and that people should put their minds at ease and devote themselves to their duties.[50]

A student of Tōzan wrote the second half of *Thoughts on Earthquakes,* which begins by explaining that many false theories continue to circulate, aftershocks (*shōdō*) have not ceased, and people fear that another large earthquake might soon occur. He points out that large earthquakes have been recorded in China and Japan from ancient times to close to the present. Therefore, earthquakes are expected occurrences, portending no future harm. Moreover, most of the current destruction has been damage to property such as storehouses, with relatively little loss of life.[51]

The substantive discussion includes observations about ground motion, the center and periphery of earthquakes, and additional discussion of indications that an earthquake is about to occur. The author first states that the earth is 90,000 Chinese *ri* in circumference, or 15,000 Japanese *ri*.[52] The distance from the center of the earth to the surface is approximately 2,500 Japanese *ri,* and the current earthquake shook an area only 200 *ri* across. The areas of shaking in any earthquake are actually quite small relative to the vast size of the earth. The author cites *Astronomy Questions Answered* for the basic idea but supplements it with additional observations.

Next, he explains that earthquakes have a center (*shin*), where the shaking is most severe. The shaking propagates outward from the center in all directions, getting weaker as a function of distance. In the case of the current earthquake, Kyoto was the center, and the outer edges were Musashi in the east, Kii in the south, Echizen in the north, and Chūgoku and Shikoku to the west. The earthquake did not come from the west or east but radiated outward from the center.[53] One reason for this observation is that people reported seeing smoke rise from the center of the city just as the earthquake began.[54] Materials produced in the aftermath of the Ansei Edo earthquake often drew on *Thoughts on Earthquakes,* with or without explicit attribution, and the material on earthquake centers was especially influential. *Ansei Record,* for example, reproduced the entire discussion, including the diagram.[55] *Ansei Record* made repeated reference to *Thoughts on Earthquakes,* as did *Ansei Chronicle* (*Ansei kenmonshi*) and catfish prints.

The discussion of earthquake-related lore is similar to that of the first part of the book. The initial topic is the original name for earthquakes in Japan and the notion of earthquake-related creatures, including a full-page illustration of an Ise Almanac (Isegoyomi) featuring the earthquake "insect" encircling Japan.[56] Toward the end of the book, the author revisits previous theories of earthquakes. For example, he takes up the idea of wind as a possible cause of earthquakes and quotes from Nishikawa Joken's *Strange Phenomena Explained* regarding this matter.[57] An afterword written by another student in classical Chinese reiterates many of the previous points: that in the past, major destructive earthquakes have been so powerful as to alter the landscape; that currently, Kyoto is experiencing many unnerving aftershocks (*yoshin*); and that there are specific warning signs of an earthquake.[58] The ultimate message, of course, is that the worst is past, and there is no need for panic or fear of what might occur in the near future.

Although written mainly to promote calm, *Thoughts on Earthquakes* also advanced seismological knowledge, particularly with respect to the idea of an earthquake center. The diagram features a circle representing the earth (fig. 3). *Jishin* (earth + center), a homophone for earthquake, indicates the center of the earth. Directly above it, a small circle at the surface indicates the epicenter of an earthquake. Two small dots, one on each side of the epicenter represent the range of shaking.[59] The phenomenon of aftershocks, in this case called *yoshin,* the same term that is used today, received extensive attention. We have seen that earlier authors were aware of such phenomena as liquefaction and differences in ground motion depending on the soil base. With respect to causes of earthquakes, however, we see find no new ideas in *Thoughts on Earthquakes* compared with works from the early eighteenth century. By 1830, yin-yang based explanations of earthquake mechanics had overwhelmed alternative theories.[60]

The Kyoto earthquake also produced a small book called *Record of Earthquakes in Japan* (*Honchō jishinki*), likewise written with the explicit purpose of calming the fears of both local residents and people in distant parts of Japan who might have been alarmed by the lurid press reports of massive destruction in the imperial capital. *Record of Earthquakes in Japan* includes a brief explanation of the causes and warning signs of earthquakes.[61] The work begins with an interesting overview of geography: "The body of the earth is yang in the north and yin in the south. Most mountain ranges are in the north. The body of heaven is yang in the south and yin in the north. Therefore, the sun moves toward the south. This is the overall pattern of heaven and earth moving relative to each other."[62] Next is the familiar explanation of the cause of earthquakes: yang energy seeking to rise but trapped by yin. Moreover, the earth is like a bees' nest, full of holes that allow fire and water to come into contact. The subsequent discussion includes a list of earthquake precursors such as well water becoming muddy and mountains appearing closer.[63] In short, there is nothing distinctive about the explanation of earthquakes or their warning signs in this work.

What distinguishes *Record of Earthquakes in Japan* from *Thoughts on Earthquakes* and other discussions of earthquakes is the inclusion of a comprehensive list of the recorded earthquakes in Japan to date, in chronological order. The earliest example is an "earthquake followed by rain" in 645.[64] Hashimoto Manpei regards *Record of Earthquakes in Japan* as the

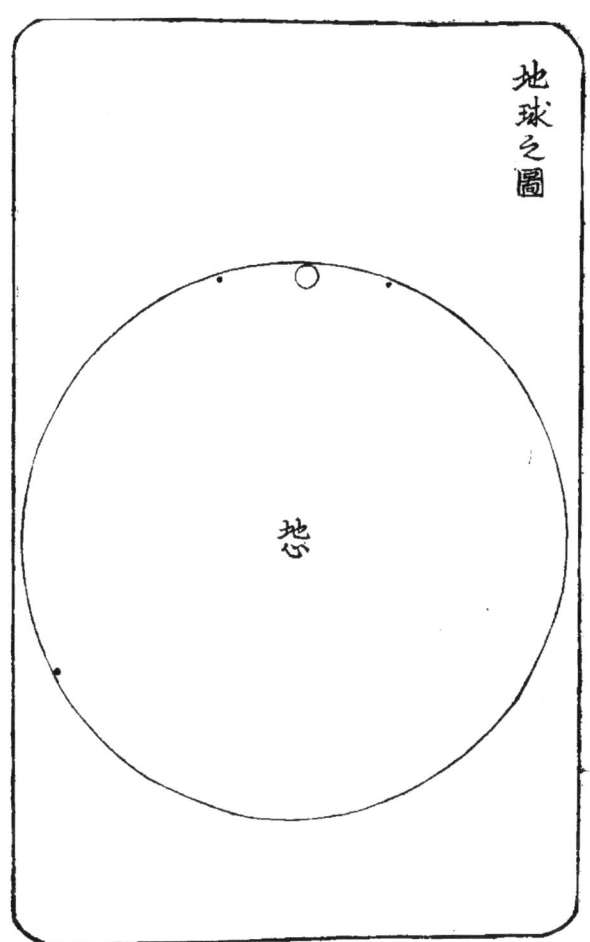

FIGURE 3
Diagram of the earth's center, the epicenter of an earthquake (top small circle), and the range of shaking (two black dots), from *Thoughts on Earthquakes* (*Jishinkō*, 1830).

earliest earthquake history in Japan,[65] and it was indeed comprehensive. As we have seen, however, nearly all discussions of earthquakes from the seventeenth century onward provided select chronologies of major past earthquakes. *Record of Earthquakes in Japan* is distinctive only in the thoroughness of its compilation.

Main Shocks and Aftershocks

As with all the large earthquakes of the Tokugawa period, diaries, letters, and journalistic accounts described not only the destruction but also other significant details about the earthquake. Aftershocks received much attention owing in large part to their psychological impact. After the main shock of the Kanbun earthquake, some people in Kyoto lived in temporary huts because "the earthquake(s) would not stop," according to a diary entry for the first day of the fifth month. An entry on the twenty-seventh day mentions, "For the past fifteen days earthquakes in Kyoto have been numerous."[66] Another account explains, "Until the sixth month, there has been some shaking every day."[67] Although the term *jishin* (earthquake) used in these and other passages clearly refers to aftershocks, it is uncertain to what extent these 1662 writers were aware of distinctions between main shocks and aftershocks.[68]

In 1830, strong aftershocks persisted for months after the main shock in Kyoto, keeping the city's residents on edge and creating an ideal atmosphere for the spread of rumors. Diaries, letters, and other accounts of the days following the main shock frequently mention aftershocks. One chronicle records one or more aftershocks almost every day for three months after the main shock.[69] "The shaking did not stop for consecutive days," according to another very typical report.[70] Another stated that "it is exceedingly unusual in Kyoto that there should be shaking from time to time all the way until the ninth month, which is why I have recorded it here."[71] A letter from two days after the main shock mentions rumors about a second earthquake and about massive fires. It then explains, "Such rumors, spread amidst the aftershocks, increase people's unease all the more."[72] Owing in large part to the explanation of aftershocks in *Thoughts on Earthquakes,* by the middle of the nineteenth century they had become widely known and sometimes appeared as diagrammatic data.

The 1856 *Ansei Chronicle* presented a method for recording the relative severity of ground movement in a series of aftershocks by using larger or smaller circles drawn under a particular day and time (fig. 4).[73] The same system appears in at least two other Ansei Edo sources.[74] Although commonly regarded as having originated in 1855 or 1856, this system can be found earlier. According to Hashimoto Manpei, there was a small earthquake in 1812 that shook Edo enough to cause minor damage to structures. A resident of Edo later received a letter from an acquaintance in Kyoto who apparently had been in Edo during the 1812 earthquake. The letter contained a drawing of different-sized circles representing the 1812 earthquake and a "Kyoto earthquake," probably from 1830. Hashimoto speculates that this letter may have been the origin of *Ansei Chronicle*'s orderly array of circles describing the relative strength of the main shock and aftershocks.[75]

FIGURE 4 Diagram of relative intensity of aftershocks, from *Ansei Chronicle (Ansei kenmonshi,* 1856). Dark circles indicate nighttime.

There is another possibility that Hashimoto does not mention. Kamahara Dōzan (read "Kanbara Tōzan" in some sources) was a samurai in the Matsushiro domain (Nagano) who compiled a large collection of materials pertaining to the 1847 Zenkōji earthquake that spans the entire year. He also authored a record of aftershocks using circles in a manner almost identical to that found later in *Ansei Chronicle.*[76] Whether Dōzan's system was entirely his own creation is doubtful in light of Hashimoto's point. In any case, however, it is clear that the aftershock recording system in *Ansei Chronicle* was not original. The same system was in use in 1847, and prior iterations may have existed since early in the century.

Precursors and Atmospheric Phenomena

Portraying earthquakes as events with explainable mechanical causes was part of the strategy in *Thoughts on Earthquakes* for promoting calm. Significantly, the book extends this approach to suggest that earthquakes might be predictable. The imbalance of yin and yang beneath the earth might manifest itself in a variety of precursor phenomena (*zenchō*). Although claims of earthquake precursors were not new in 1830, *Thoughts on Earthquakes* was the earliest widely read book to compile these alleged precursors with an eye to the future. We saw that Nishikawa Joken expressed skepticism regarding the possibility of earthquake prediction by divination, and indeed no influential Japanese works on earthquakes regarded them as amenable to divination in advance. The new idea that earthquake precursors exist and might be useful in predicting a main shock derived from yin-yang explanations of earthquakes. Starting in the nineteenth century, certain phenomena that suggested excess heat or emissions of hot vapor from within the earth came to be routinely listed as precursory phenomena. Examples include unseasonably warm weather, steam or hot mist, stars or the moon appearing red, peculiar clouds in the sky, changes in well water, emissions of light, and (especially after 1855) strange animal behavior. Significantly, contemporary earthquake prediction in Japan and some other parts of the world is based on the same basic claim that we find in *Thoughts on Earthquakes* and other nineteenth-century works: if only we can catalogue and study a sufficient variety of precursors, we might be able to predict earthquakes. Moreover, many of the specific precursory phenomena claimed by contemporary advocates of this approach are the same phenomena found in nineteenth-century texts. I examine this topic in detail in chapter 6. The purpose of this section is to survey the major varieties of early modern precursors and to take note of their theoretical basis, a point often ignored by modern and contemporary advocates of the value of these phenomena for earthquake prediction.

Light Flashes, Thunder, Lightning

Perhaps more than any phenomenon other than shaking, Japanese lore associated earthquakes with flashes of light. Often this association included thunder and lightning, which many early modern books regarded as essentially the same phenomenon as earthquakes but occurring above the ground. Nineteenth-century accounts of earthquakes usually included a report on weather conditions, and sure enough, clouds and thunder were usually present prior to the shaking.

Reports of light flashes seem to have begun in 1662. Asai Ryōi described "shining objects" (*hikarimono*) in the skies that traveled to the top of Mt. Hiei, which were witnessed by many during the earthquake.[77] Other documents from the Kanbun earthquake mentioned shining objects, and the *Konoe Diary* (*Konoe nikki*) even included a small line drawing of such an object, resembling a spoon or ladle, or possibly a comet.[78] From that time to the present day, flashes, objects, or pillars of light have become common elements in Japanese descriptions of earthquakes. Moreover, in modern times, "earthquake lights" have been widely reported throughout the globe. Despite many hypotheses about these phenomena, no one has been able fully to describe these lights, much less determine their cause.[79] In early modern Japan, the likely reason for the association of earthquakes and light flashes was the emerging consensus about the mechanical cause of earthquakes, which regarded them as underground thunder caused by explosive upwelling of yang energy. Thunder, of course, was closely connected with lightning. In this way, earthquakes and rainstorms also became associated in the popular imagination. *Foundation Stone* was not the origin of the association of earthquakes and light flashes, but it helped to popularize the connection.

During the Genroku earthquake, "flashes of light that resembled lightning" moving toward the southeast could be seen after the earthquake, and the earth emitted a roar, "like thunder" just before the earth shook.[80] Similarly, 152 years later, the Ansei Edo earthquake began with a light show according to many accounts. "In the east, a flash of light appeared, similar to lightning. A moment later it was gone."[81] "As soon as a suspicious shining object flashed across every direction, the earth groaned."[82] In some versions, the light flashes emanated from or moved across the night sky: "In the southeast of Edo, a shining object resembling fire crossed the sky."[83] In other descriptions, the light came from the ground up: "At the time of the present earthquake, fire ki emanated from within the earth."[84] Workers on ships in the harbor reported seeing flashes of light over Edo at the time of the earthquake.[85] Particularly spectacular is the account in *Seeds of Tales from the Great Edo Earthquake of the Latter Age* (*Edo Ōjishin matsudai hanashi no tane*), an anonymous, sensational book from 1855. It tells of a series of light flashes, "white ki," that emanated from fissures in the ground from the Nihon bank of Shin-Yoshiwara to Sensōji. This white ki struck the nine-wheel spire (*kurin*) atop the temple's five-story pagoda, bending it before dispersing in all directions. *Seeds of Tales from the Great Edo Earthquake of the Latter Age* includes a dramatic image of a massive beam of light emerging from the ground at a forty-five-degree angle and causing an explosion above the spire.[86] *Fujiokaya Diary* (*Fujiokaya nikki*) explained that a "shining object" traveling across the sky from north to south caused the spire to bend.[87] The bent spire became an iconographic emblem of the earthquake, appearing in popular broadsides, catfish prints, journalistic literature such as *Ansei Chronicle,* and even in a newspaper article from the Meiji period.[88]

Dramatic emissions of light were reported widely in 1896 and 1933 in connection with the massive earthquakes and tsunamis off the Sanriku coast in northeast Japan. Meiji Sanriku earthquake witnesses also frequently reported booming cannonlike or thunderlike noises. Shōwa Sanriku earthquake investigators noted lightinglike flashes of light reported throughout the region.[89] Although seismologists have never been able to identify plausible sources for light flashes or related phenomena, it was not for lack of trying. Writing in the 1950s, Musha Kinkichi, for example, defiantly urged students of earthquakes to take these descriptions of flashes of light seriously.[90] Not long after Musha's time, the increasing acceptance of plate tectonics may have helped reduce interest in such phenomena, at least among seismologists. For example, Miki Haruo, in a late 1970s book on the Kyoto earthquake, gives only passing mention of light flashes and concludes that "the identity of the shining objects has been unknown from then to now." Later, he concludes that there is no statistical relationship between earthquakes and rainstorms.[91]

As we have seen, however, reports of and interest in earthquake lights continue to the present. The sheer volume of such reports is the main reason many seismologists will not dismiss the phenomenon outright. Earthquake lights, strange animal behavior, and other elusive possible precursors are typically identified in retrospect, after the earthquake has struck. I would argue that it is important to bear in mind the psychological dimension of these observations. Stated simply, people tend to see—or to believe they saw—what they expect to see.[92] The early modern understanding of earthquakes as an underground version of thunder and lightning was especially conducive to perceiving flashes of light when the earth shook.

Other Precursors

The 1828 Sanjō earthquake produced a lengthy work on earthquakes, Koizumi Kimei's *Account of Chastisement and Shaking* (*Chōshin hiroku*). It begins with a detailed description of the geography of "our Echigo country" and then discusses precursors of earthquakes. For example, from about the seventh or eighth day, a thick foglike substance appeared in the mornings that obscured people's view, and a five-color rainbow appeared around the sun. A sound like thunder occurred just before the earth shook. The ground in fields moved in a wavelike manner when the shaking began, and the earth tore apart in places. The earthquake caused mountains to collapse and expelled muddy water from wells. It also expelled fire and fiery wind from within the earth, a reflection of the idea that yang energy under the ground causes earthquakes.[93]

Less than two years later, *Thoughts on Earthquakes* argued that because earthquakes are predictable with careful attention to precursors, in the absence of specific precursors people need not worry that a new main shock will soon strike. The main indications of earthquakes are dirt issuing forth from small holes in the ground (like the activity of moles), smoke coming from fields when they are plowed, and the water in wells becoming muddy.[94] Mention of smoke rising from the center of Kyoto in the second half of *Thoughts on Earthquakes* segues to further discussion of precursors. Examples include the sun and moon shining an abnormally red color and resembling a dish in the morning and evening. An observer reported mountains taking on a strange appearance in terms of their color just before the onset of shaking. In another case, a mineshaft conducted large quantities of steam or smoke—"earth ki"—upward, obstructing everyone's view from the waist up. Knowing it was a sign of an earthquake, nobody went into the mine and all escaped unscathed. Just before the shaking started, several thousand heron all took flight at once, because birds can also detect upwelling of earthly ki. Another warning sign was the appearance of a rainbow in places or circumstances where it would not usually be seen. These alleged warning signs came not from the author's direct observations but from reports and written accounts.[95]

The 1847 Zenkōji earthquake became especially prominent in the discourse following the Ansei Edo earthquake because the people of Edo retrospectively associated the two events. In hindsight, the Zenkōji earthquake became a source of potentially valuable information about earthquake precursors. In one account, a man who had experienced the Zenkōji earthquake told his son of certain cloud formations that had appeared before it began. Seeing the same clouds in the sky over Edo in 1855, the son removed valuables from his house, placing them in an open area, and took other precautions just in time.[96] Another piece of useful information from Zenkōji was that water levels in wells decreased prior to the earthquake.[97] The idea that changes in wells could predict earthquakes became a prominent feature of earthquake discussion in modern times.

One tale that circulated right after the Ansei Edo earthquake was that a giant magnetic stone approximately one meter in width, located at a shop called Nanigashi that sold eyeglasses, lost its magnetic properties roughly two hours before the main shock. Reported in *Ansei Chronicle* and *Fujiokaya Diary,* this tale is impossible to verify. Modern authors promoting earthquake prediction sometimes cite this 1855 "fact" as evidence supporting the hypothesis that electromagnetic anomalies often precede earthquakes.[98] In 1855, the tale spread widely and resulted in the design of several warning devices. Most famous is a sketch of an earthquake alarm clock (*jishinkei*) that theoretically might provide brief advance warning. A magnet held a small metal weight above the ground. Soon before an earthquake struck, the magnet would lose its power, thus releasing the weight, which was attached to a string. The string wound around a wheel that controlled a metal ringer. If all went well, the bell would ring several times, giving anyone in hearing range warning to prepare for an earthquake.[99] There is no evidence that the device advanced beyond the drawing stage, despite claims by some modern writers that it existed.[100] Sakuma Shōzan (1811–1864) produced a prototype of a simpler device based on the same principle. The device itself is not extant, but a photograph Shōzan made of it is. It consists of a horseshoe-shaped magnet suspended from a string, to which a metal nail weighing about ten grams is attached at the bottom.[101] A sketch of a different device appears in Murayama Masataka's 1856 *Thoughts on Earthquakes and Electricity* (*Shinden kōsetsu*). It consists of a magnetic stone roughly six inches in diameter held above a basin by a stem. Various nails and pins hang from the bottom of the rock and should fall into the basin to make a warning sound before an earthquake strikes. Masataka's explanation reads, "In our country, from ancient times to the Ansei era, there have been instances in which earthquakes were connected with electricity [or thunder and lightning]. The diagram here describes an earthquake warning device based on a magnetic stone."[102]

These devices made good sense based on the theory that magnets lose their magnetic properties just before an earthquake strikes. Because their theoretical basis was inaccurate and derived from unconfirmed rumor, these devices represent an early dead end for earthquake prediction research.

Scientists such as Shōzan and Masataka were undoubtedly aware that electricity played a major role in many European theories of earthquakes, and general connections between electricity and magnetism were well known by this time. This background knowledge probably added credence to the tale from 1855 concerning the loss of magnetism. Furthermore, Japanese authors often did not distinguish between electricity in general and specific manifestations such as lightning, which had long been associated with earthquakes. A variation on this basic idea, one that continues even today to receive serious attention, is that certain fish can predict earthquakes. One speculative reason that some scientists posited during the twentieth century was that some fish are unusually sensitive to electrical signals given off just before an earthquake. The origins of the idea of fish as earthquake predictors also come from the Ansei Edo earthquake, and I discuss modern fish research and speculation in detail in chapter 6.

As we have seen, catfish became increasingly well-known symbols of earthquakes during the eighteenth century. Just because catfish were symbols of earthquakes, however, did not necessarily mean that anyone regarded them as predictors of earthquakes. Works such as *Thoughts on Earthquakes* list a wide variety of phenomena as possible indicators that an earthquake is about to strike but say nothing about catfish or any other species of fish. As Hashimoto points out, it was only after the Ansei Edo earthquake that some writers credited catfish with earthquake prediction.[103] The apparent locus of this idea is the 1856 *Ansei Chronicle*. Although the author is unknown, most of the text was probably written by Kanagaki Robun. It contains an account of an eel fisherman named Shinozaki from Nagakura-chō in Honjo who tried various spots along the river on the evening of the second day of the tenth month. He caught not a single eel, but catfish were unusually active. He caught three catfish and recalled that when catfish are agitated and active, an earthquake will soon strike. Shinozaki returned home, spread a mat outside, put all his possessions on it, and otherwise made emergency preparations. His wife thought his actions ridiculous, but that night the earthquake struck. Another fisherman nearby Shinozaki ignored the message of the catfish agitation and continued to fish. He returned home to find his dwelling and possessions destroyed. The passage concludes by moralizing about the virtue of Shinozaki's perception and states that people who knew of the story realized "that negligence is one's own fault." *Ansei Chronicle* characterizes the agitation of catfish as the earth starts to move as "a natural principle."[104]

The specific idea that catfish can predict earthquakes seems to have started in 1855 or 1856, but there is some precedent for a general notion in folklore of certain fish as transmitters of messages. Miyata Noboru points out that a common motif in Japanese folklore along bodies of water is that catfish, along with eels, are "fish that can transmit messages" (*mono iu sakana*). In such lore, the fish are able to speak or turn into human form to transmit their messages. Catfish and similar fish appeared in tales in which people, typically fishermen, would excessively impinge on their domain— for example, a fisherman trying to catch a catfish with a cormorant. The catfish would assume human form and try to pass on a warning to stop the impingement. If the attempt failed and the fish was killed, because it was the ruler of the realm of water, some curse or punishment would come forth from heaven, often a flood.[105] The circa 1855 transformation of catfish into predictors of seismic events was likely a variation of this general motif. The prominence of catfish in the cultural products associated with the 1855 earthquake surely aided in this transformation.

The Mature Early Modern View

Bukō Chronicle (*Bukō nenpyō*) is a record of major historical events, mainly in Edo, compiled by Saitō Gesshin (1804–1878) in 1848 with some additions in 1878. Gesshin was an academically inclined neighborhood head in Edo. His account of the 1703 Genroku earthquake is a good example of an educated person's understanding of earthquakes in mid-nineteenth-century Japan. Gesshin describes strong thunder during the evening prior to the earth roaring "like thunder" at the eighth hour that night. He describes houses undulating like boats and fissures in the ground as large as six *shaku* (roughly six feet) that blew out sand or water. Soon after the eighth hour, four tsunami waves swept in and out of the rivers. Aftershocks continued frequently for about a week. Following this description, Gesshin includes a line of verse from the imperial court poet Naka-no-in Michishige (1631–1710): "The rock of the thousand generations of the country's *kami* will not be shaken down by pulling on the immovable August Reign."[106] As we have seen, verses emphasizing the stability of the country's foundations, usually in the form of *kami* and the imperial court, were common in the wake of major earthquakes. Of course, Gesshin summarized this earthquake from documents, and the purpose here is to note the language and concepts he employed from his circa 1848 vantage point.

Rain Dampened Sleeves (Shigure no sode) is an account of events and lore connected with the Ansei Edo earthquake compiled by Hata Ginkei. The chapter discussing earthquakes provides a useful glimpse of the knowledge available to well-read Japanese around 1855. It consists mainly of a summary of different theories and topics, but Ginkei's own ideas also guide the discussion. He discusses wind as a source of the trapped yang energy that causes earthquakes and points out that fierce wind accompanied an earthquake in the Kantō area in 1648. Subsequent sections include quotes from *Illustrated Compendium of Chinese and Japanese Knowledge,* a Chinese astronomy text, *Chronicles of Japan (Nihongi* or *Nihonshoki),* and several passages from *Thoughts on Earthquakes.* Ginkei includes a diagram of an earthquake's epicenter, extensive discussion of signs of an impending earthquake, and a discussion of earthquake divination.[107] *Rain Dampened Sleeves* does not present new theories, but Ginkei assembles and summarizes in one place nearly all of the accumulated knowledge about earthquakes to date. Ginkei also provides an interesting explanation of how catfish came to be associated with earthquakes. From far back in the past, children imagined a giant catfish to live under the earth and cause earthquakes when it moved its whiskers *(hige).* Although the catfish idea is "utter nonsense," the likely explanation for it is that the *na* in *namazu* (catfish) is the same *na* as in *nai* (earthquake). Furthermore, speculates Ginkei, *hige* (whisker, beard) and *hire* (fin) became confused. Therefore, catfish whiskers became associated in children's lore with earthquakes.[108]

Ansei Record was a widely read journalistic account of the Ansei Edo earthquake. Its explanation of earthquakes, which appears in the initial part of the section "Jishin no ben" (Earthquakes explained), should be tediously familiar by now. It begins with the *Astronomy Questions Answered*–derived explanation that heaven and earth are full of holes like a bees' nest or the cap of a mushroom. The basic mechanism is that fire energy seeks to rise but is blocked by water energy—the same logic as thunder in the heavens. Moreover, *Ansei Record* provides exactly the same geographical explanation as previous texts regarding the frequency of earthquakes with respect to the North Pole, the equator, and warm areas with rocky soil. One point about the causal mechanism of earthquakes is an explanation that although Edo experiences frequent small earthquakes, the relative absence of large ones is because of the presence of many wells. These wells ordinarily provide yang energy with an escape route.[109] This point was important, as we will see.

By 1855, the basic idea of yang energy within the earth seeking to rise but blocked by yin had long been the dominant commonsense mechanical cause of earthquakes. Moreover, well-read Japanese of this time with an interest in natural science would likely have been aware of aftershocks, liquefaction, sand and mud blows, uplift, epicenters, and subsidence.[110] They also had a vague sense of hypocenters and seismic waves, the knowledge that shaking diminishes farther from the epicenter, and that the severity of ground motion varies as a function of the soil base. Furthermore, tsunamis and volcanism were both understood as closely connected with earthquakes. In short, careful observation within an increasingly literate society had established a solid foundation of geophysical knowledge. The major gap in this knowledge, of course, was an understanding of faults and plate tectonics, two concepts that would not become known and accepted anywhere in the world until the early and later twentieth century respectively. In the absence of this knowledge, yin-yang–based descriptions of earthquakes had great explanatory power.

Other important components of the late Tokugawa-period view of earthquakes included an assumption that earthquakes and thunder were both the same basic phenomenon. Moreover, in academic circles there was an increasing interest in the possibility that electricity might be the key to explaining earthquakes. Most Japanese at this time thought earthquakes had connections with atmospheric phenomena, especially wind. Smoke, steam, other emissions of heat from the ground, unseasonably warm weather, and brightly glowing stars and the moon, especially with a reddish tint to them, were all signs of earthquakes, albeit always recognized after the fact. The basic idea behind all of these supposed precursors was yang energy rising from the earth. If this energy could be vented to the surface effectively, all would be well. If it became bottled up, explosions and shaking, accompanied by the expected sound and light effects, were the likely results.

The Ansei Edo Earthquake as a Turning Point

According to prevailing wisdom at that time, the Ansei Edo earthquake should never have happened because water wells were common throughout the city. The possibility of a large earthquake was therefore beyond the imagination of Edo's residents, thus adding to its psychological impact. What about the Genroku earthquake of 1703? Why did that event not suggest to the people of 1855 that Edo was subject to large earthquakes?

Historical amnesia may have played a role, but the main factor was the history of Edo's water supply. In 1703, Edo's water supply was mainly a series of open pipes that brought water in from upstream areas of the major rivers. Three major networks—the Kanda system, the Tama system, and the Mita system—supplied water of adequate quality to all of the warrior housing areas and most commoner areas. Starting in the eighteenth century, however, many daimyō and other samurai began digging their own wells to obtain better and more convenient water. Gradually, commoner neighborhoods also began digging wells, and communal wells became the center of neighborhood micro communities.[111] This building of wells began just after the Genroku earthquake, so it was easy to explain that event as a buildup of yang energy within the earth but also to think that future large earthquakes were no longer possible.

Anyone who gave serious thought to the Ansei Edo earthquake's physical causes, therefore, would have had to question the theory of trapped yang energy. One reaction was to turn to Dutch books in search of better theories, or at least alternative ones. Moreover, as exemplified by designs for earthquake warning devices, scientific literature after Ansei Edo reflected optimism that ways of predicting or defending against earthquakes could be found. One particularly interesting example is Utagawa Kōsai's 1856 *Earthquake Prevention Theory* (*Jishin yobōsetsu*). Its main text was based on a recent translation of an article in a Dutch periodical, *Nederlandsch magazijn*. *Earthquake Prevention Theory* dismissed a variety of earthquake explanations such as cave-ins, underground explosions of combustible material, or the violent upwelling of steam or hydrogen gas. "There is but one source of earthquakes, and it is electrical [*erekiteru*] power. . . . Earthquakes come about because of electricity [*denki*] lying under the earth." Just as electricity in clouds produces lightning, a similar process occurring under the earth produces an earthquake. The text concludes by arguing for the need to place the equivalent of lightning rods within the earth. Metal rods inserted deep into the earth would conduct electricity up, out, and into the air. Digging the deep holes for this purpose would be expensive, which is why no government or other entity had yet put it into practice.[112]

This electrical theory had a long history in Europe, and it fit in well with observed and imagined phenomena in Japan.[113] It was similar to the trapped yang theory but sufficiently different to explain the 1855 event. Moreover, it fit well with the idea of earthquakes as underground thunder and lightning and with the many reports of energy or light emanating from the ground during earthquakes. Electricity was very much in the intellectual air in 1856. Another work published that year, Yamazaki Bise's *Thoughts on the Year of the Great Earthquake* (*Ōjishin rekinen kō*), discusses earthquakes, thunder, and lightning at some length. It then provides what might be the earliest example of explicit instructions for constructing an earthquake-resistant building.

Because electricity was the force behind both earthquakes and lightning, the proposed design model would defend against both (fig. 5). A wooden pole in the center of the structure, sunk into the earth, wrapped in a stone base, and protruding well above the roof, would serve as the base for iron extensions attached to its top. Four large jars were sunk into the earth like wells, one on each side of the structure. An iron chain attached the top extensions of the central lightning rod to each well. The system would conduct electricity from storms or within the earth to the top of the lightning rod. Presumably, it would dissipate into the air or into the pole made of nonconducting materials.[114] There is no evidence that anyone attempted this construction technique, and it raises the question of how the dissipation of electricity on such a small scale could actually reduce ground motion. Nevertheless, its appearance at this time is an early manifestation of an active approach to mitigating earthquake hazards, even if utterly ineffective.

Two other translations of Dutch works on natural science contain explanations of earthquakes. Kawamoto Kōmin's *Lectures on Natural Science* (*Kikaikanran kōgi*) is a translation of Johannes Buijs' (1764–1838) *Natuurkundig schoolboek*. The translation took place between 1851 and 1857, with the explanation of earthquakes appearing in either 1856 or 1857. It puts forth two theories. One is that hydrogen, when heated, combines with oxygen, undergoes combustion, and causes the earth's crust to move. The second is that saltpeter, charcoal, and sulfur are located in close proximity at the focus of an earthquake. Heated oxygen ignites this mixture, causing combustion, ground motion, and noise. Similarly, Hirose Genkyō's *Essentials of Natural Science* (*Rigaku teiyō*) is based on his reading of several Dutch books. His explanation of earthquakes relies on the basic idea of explosions within the earth causing ground motion. When this phenomenon manifests itself above the earth, it is in the form of volcanic activity. Below the earth, it becomes an earthquake. The basic mechanism is for saltpeter, charcoal, and sulfur within the earth to explode. There are three possible ignition sources: (1) rocks that emit fire; (2) hydrogen within the earth that mixes with oxygen in the atmosphere to create a combustible combination; and (3) rocks or soil from above that fall into holes and crevices in the earth, generating fire on impact.[115]

These new ideas were not highly influential, but their appearance in 1856 and 1857 indicates that the Ansei Edo earthquake encouraged a questioning of received knowledge in academic circles.[116] Indeed, a major significance of that earthquake is that it prompted a search for alternatives to the yin-yang theory of earthquakes. Moreover, the plans for earthquake prediction devices and earthquake-resistant building design indicate a new attitude toward earthquakes. The general idea was that careful observation

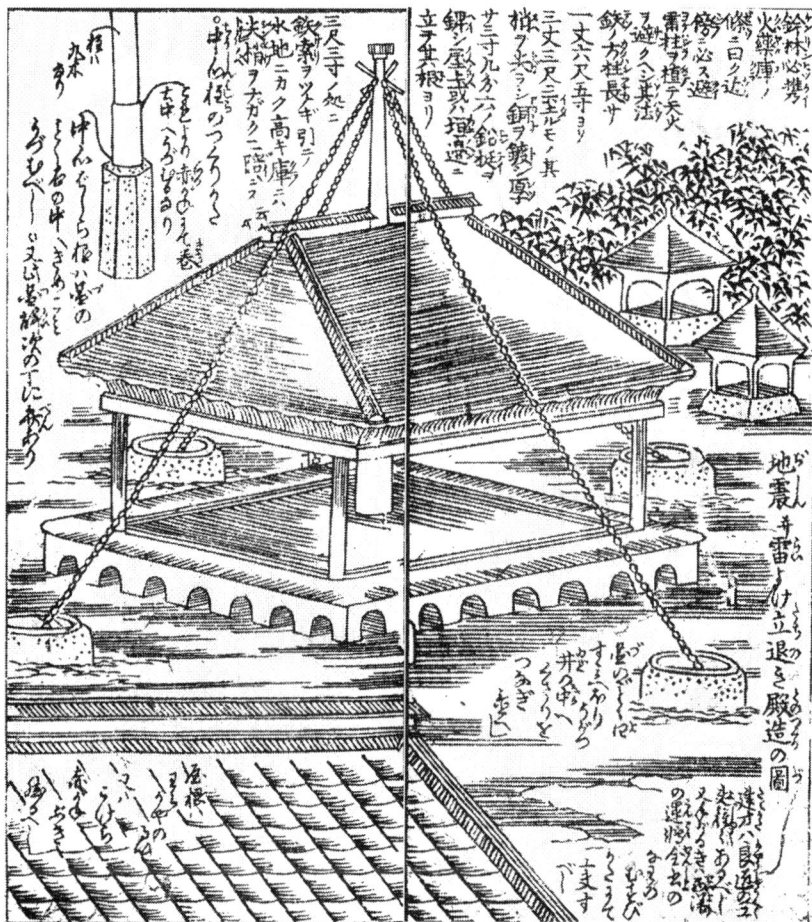

FIGURE 5 Perhaps the earliest example of earthquake-resistant building plans in Japan, from *Thoughts on the Year of the Great Earthquake* (*Ōjishin rekinen kō*, 1856).

of seismic precursors and a better understanding of the mechanics of earthquakes could serve as the basis for earthquake prediction and engineering innovations to mitigate earthquake damage. In chapter 6, I explain that this notion helped transform the Ansei Edo earthquake from an event characterized by renewal in 1855 and 1856 into a menacing example of society's failure to heed the warning signs during the early Meiji period. More recently, events such as the Hanshin-Awaji (Kobe) earthquake of 1995 and of course the Great East Japan Earthquake of March 11, 2011, remind us that we still lack any practically useful warning signs to read, despite folk claims to the contrary. Although earthquake prediction in any useful sense remains elusive, earthquake-resistant building and infrastructure design has proven to be highly effective.

CHAPTER 3

Japan according to Earthquakes

Early modern earthquakes shook complex societies. Prevailing perceptions of the natural world, religious concepts, moral norms, political practices, and other resources were available as tools for making sense of major earthquakes. These varied social resources, however, were often insufficient to explain destructive events of such great magnitude in an entirely satisfactory way. Consequently, earthquakes could function as catalysts, accelerating social changes already under way. In some cases, they even stimulated new developments. One example is the Ansei Edo earthquake stimulating new ideas about the mechanics of earthquakes and therefore how to predict and defend against them.

Major earthquakes produced thought and rhetoric about religion, morality, politics, and sometimes the very foundations of society. Time, place, and circumstances mattered. For example, the development of a nationwide network of news distribution and substantial reading audiences in urban areas by 1830, combined with an earthquake that shook the imperial capital during a year of special religious significance, helped create sensationally exaggerated media reports. The fear such reports generated was in turn an impetus for the publication of *Thoughts on Earthquakes* (*Jishinkō*) and other works encouraging a calm, rational approach to earthquakes. *Thoughts on Earthquakes* advanced scientific knowledge and helped condition the way Japanese in the 1840s and 1850s reacted to earthquakes.

Earthquakes functioned both as agents of social change and as windows affording a particularly clear view of society. The previous chapter explains the development of a dominant view concerning the mechanical causes of the earth's bouts of violent shaking. This chapter moves the analysis into the realm of the social effects of earthquakes. I argue that earthquakes contributed to shaping the social and imaginative contours of Japan. Moreover, earthquakes even created some of the physical contours of Japan. The claim in *Ways of Earthquakes in Japan* (*Honchō jishin no shidai*) that Lake Biwa and Mt. Fuji arose from an ancient earthquake was, of course, an oversimplification of complex geophysical forces. It was a plausible idea, however, because by 1855 the phenomenon of earthquakes creating land, lakes, and mountains had become well documented.

Religion

Despite widespread agreement on the basic physical mechanisms that produced earthquakes, these mechanisms were only the proximate cause of shaking. At least for some Japanese, social factors such as moral corruption, extreme imbalances in wealth, or government malfeasance might have a role to play in earthquakes or other disasters. A lack of clear boundaries between natural and social phenomena encouraged a tendency to view these realms as interconnected. Consider, for example, these words by Kaibara Ekken (1630–1714), which display typical Confucian macrocosmic-microcosmic thinking: "If the flow of material force (ki) through heaven and earth is obstructed, abnormalities arise, causing natural disasters such as violent windstorms, floods and droughts, and earthquakes. If the things of the world are long collected together, such obstruction is inevitable. In humans, if the blood, vital essence (ki), food and drink do not circulate and flow, the result is disease. Likewise, if vast material wealth is collected in one place and not permitted to benefit and enrich others, disaster will strike later."[1] In just a few sentences, Ekken has linked the human body with the forces of nature and the economic well-being of society.

Nearly two centuries later, the catfish print *Earthquake Fortune Explained* (*Jishin kikkyō no ben*) expressed a similar point that everything from human anatomy to cosmic anatomy is interconnected. The lengthy text of the print ranges widely. It begins with a pregnancy metaphor: "Earthquakes are the basis for years of bountiful harvests. What reason is there to be angry with them? In autumn the trees and grass return to the earth, and the energy of winter produces an abundance of sprouts within the earth. The blessing of Heaven makes the earth pregnant, giving birth to all things. However, things appearing in violation of the proper temporal sequence cause earthquakes. Earthquakes are the result of the labor pains of the earth."[2] This print argues, often in convoluted detail, that while a misalignment of cosmic forces and the seasons helped cause the earthquake, in the longer run the destruction of the earthquake will lead to renewal. The print also has much to say about human morality: "People who are not socialized lack gratitude. Humans regard heaven and earth as their father and mother, and all things and the people of the four seas are their siblings. On this basis, it is the ethical human way for the old to guide the young and for the young to assist the old. In recent years, however, humane feelings have grown thin. The gap between self and others is strong, and people delight in luxurious food. They take delight in out-of-season flowers, spending much money, and acting contrary to heavenly principles. For example, even those who escape earthquake damage turn their backs on these teachings, although they should be humble."[3] In short, the recent earthquake has highlighted an alleged moral degeneration of society. That this moral decay played a role in causing the earthquake is implied but not clearly articulated. Underlying the text is a vague notion of correlative cosmology in which personal behavior, human society, and cosmic forces are interconnected.

The bulk of the discussion in *Earthquake Fortune Explained* argues that the human body is a microcosm of the broader cosmos. Humans, therefore, are bound by cosmically ordained ethical rules. Discarding or ignoring these rules can result in such corrective action as the current earthquake: "Indeed, because humans have received the character of heaven and earth, there is no characteristic of heaven and earth that is not reflected in people. Heaven is round, and therefore people's heads are round. The sun and the moon are in the heavens and thus people have two eyes. The stars are arrayed in the heavens and people have rows of teeth. Heaven produces wind and rain, and humans have the emotions of happiness and anxiety. Heaven speaks with thunder, humans speak with vocal sounds. Heaven possesses yin and yang, humans male and female."[4] The full discussion goes on at greater length, but one argument should be clear at this point: humans are a reflection of cosmic forces in every aspect. The implication is that the earthquake was not a random occurrence, independent of human society and behavior.

There was a range of possibilities for imagining interactions between the cosmic forces and human society. Deities, usually expressed by the stock phrase "the *kami* and the buddhas," played a major but not absolute role. The *kami* and buddhas were powerful forces. Typically, they were imagined much like local human warlords in their castles. Residing in their shrines and temples, *kami,* buddhas, and bodhisattvas watched over their territory and intervened by bestowing rewards or punishments. Resembling powerful humans, deities could be neutral, protective, or antagonistic vis-à-vis an individual or organization. Some were more powerful than others, but none was all-powerful. Only as a team, for example, might a group of deities be able to start or prevent something as large as a major earthquake. Sometimes, the convulsions of the earth were beyond their power to control. Popular prints produced in the wake of both the Odawara and Iga-Ueno earthquakes echoed this point: "Even the benevolence of the deities and the buddhas would have been unlikely to calm the situation."[5]

The power of a deity was usually greatest in the area near its home base. The deities were not transcendent forces. They relied for their power on human society, just as human society benefited from divine assistance. The first article of the *Jōei Formulary* of 1232 states, "*Kami* increase their power by virtue of people venerating them, and people encounter good fortune by means of the divine influence of the *kami.*" At the time, this view of deities was new and stood in contrast to an older view of them as capricious and unpredictable. This *Jōei Formulary* assertion was an early example of the medieval theory of mutual interdependence between deities and humans.[6] What was a theological concept in medieval times became widely accepted common sense by the Tokugawa period.

Let us consider two examples from the 1850s that illustrate views of the deities from a popular perspective. In connection with the intimidating visits of Commodore Matthew Perry in 1853 and 1854, the bakufu began an ambitious plan to create offshore artillery batteries on artificially constructed islands in Edo Bay. The name for these batteries was *daiba* (or *o-daiba*), and they quickly became an iconic symbol of bakufu power or weakness, depending on one's point of view (fig. 6). A satirical tale circulating in 1854 described the annual meeting of Japan's major deities at Izumo. The assembled deities call on Amaterasu to summon forth a divine wind (*kamikaze*) to drive away the Americans, but Amaterasu tries to pass the buck and get Shakyamuni (the Buddha) to do the job. Amaterasu's ostensible reason is that these days, the "way of the *kami*" is in decline relative to the prosperity of Buddhism. Upon hearing Amaterasu's request that he produce a "Buddhist wind," Shakyamuni nervously declines because he has heard there are many "*daiba*" in Edo Bay. "Daiba," written with different characters, was a well-known abbreviation of Daibadatta (Sanskrit: Devadatta), an evil disciple who tried to kill Shakyamuni.[7] Though amusement was the main point of this tale, it nicely illustrates popular views of the deities. They are powerful but far from invincible. They might act as a group during a crisis, and their motives are similar to those of humans.

Similarly, the Ansei Edo earthquake occurred during the tenth lunar month, the "month of no deities" (*kannazuki* or *kaminazuki*), a time when the major deities of Japan leave their home bases and travel to Izumo for a convention.[8] While away on business, these deities leave subordinates behind to attend to local matters. Ebisu was the designated caretaker in the case of the Kashima deity (Kashima Daimyōjin), the most powerful *kami* near Edo. Almost as soon as Kashima departed, the earthquake occurred. It was common for catfish prints to portray Ebisu as incompetent and to suggest that Kashima's absence was a cause of the earthquake.

How might someone take advantage of the power of the deities? Good behavior was one obvious answer. After the 1847 Zenkōji earthquake, which resulted in thousands of Buddhist pilgrims dying in and around Zenkōji, a temple official speculated on the cause of the disaster. Was it because he had eaten fish or meat that evening, which the resident priest had been hiding? Perhaps defiling such a holy place had angered the goblins (*tengu*) or

FIGURE 6 Ruins of an offshore artillery battery (*daiba* or *o-daiba*), 2010. Photo by MachineCitizen, Wikimedia Commons.

mountain spirits?[9] It was common, incidentally, to regard local spirits and supernatural creatures as advocating Buddhist values. In any case, morally good behavior, while certainly desirable, was only a start.

For many early modern Japanese, divine protection was an investment process. Devotional acts might pay good dividends, even in the short run. In the wake of the Zenkōji earthquake, for example, a tale circulated of a twenty-year-old woman who survived in a house buried under mud for some twenty days. Fortuitous circumstances of furniture placement provided her with air space and food, and she extracted water from the mud. The main reason she survived, however, was her constant intoning of the names of the *kami* and buddhas.[10] In a more typical example from the Ansei Edo earthquake, the deity of Mt. Narita Fudō of Shimōsa saved the life of a man who relied on a protective amulet from the deity's shrine. In this case, the deity did not intercede in a dramatic manner but simply provided warnings that aroused the man's sense of caution.[11] Presumably, the deity was powerful enough to know the earthquake was imminent. The majority of earthquake tales involving religious matters tell a similar story: a devoted follower of a particular *kami*, buddha, or bodhisattva survives the earthquake, and that survival is explicitly or implicitly attributed to the deity's assistance. It is a rather straightforward payback for those who invest in a deity in advance.

Manifestations of piety in the midst of a disaster might help, but an active investment of money or some other valuable resource in advance was the best way of purchasing divine insurance. Some investments, however, did not work out. According to one account, after the 1596 Keichō-Fushimi earthquake, Toyotomi Hideyoshi visited Hōkōji to check on a statue sixteen *jō* high (approximately forty-eight meters) of Rushana Buddha that he had erected there in 1588. He became angry at seeing the badly damaged statue, specifically installed to promote the prosperity of the state. Decrying the waste of time and money, Hideyoshi said to the statue, "Your big body is a disgrace. We should not trust this broken and useless buddha who cannot even protect himself!" He then shot an arrow into it, to the shock of those nearby.[12] Sixty-six years later, following the Kanbun earthquake, *Rain and Shine Diary* (*Seiu nikki*) repeated this tale of Hideyoshi's chastisement of the statue.[13] Earthquakes were a supreme test of divine power, and not all the divinities passed.

It was common practice to seek talismanic protection amidst the stress of fires and aftershocks. After the main shock of the Kanbun earthquake, survivors copied sacred verses and affixed them to houses, gateposts, or roof beams. One verse invoked the comforting thought that we live in a sacred country (*shinkoku*) forged by the deity Izanagi. Another invoked Hachiman to protect the people.[14] Asai Ryōi's *Foundation Stone* discussed these matters and similar religious lore. In one case, unruly mobs went to the Toyokuni Shrine, which had suffered no damage. They pulled up grass from the shrine's inner precinct and hung it from the eves of their houses as protection while aftershocks continued. Local officials conducting damage surveys chastised the residents of such houses, which caused visits to the shrine to drop off. At the same time, rumors about sacred horses swept through the city. Horses from the Iwashimizu Hachiman Shrine disappeared after the shaking, returning later covered with perspiration. Likewise, the horses of the Kashima Shrine in far-off Hitachi Province became agitated, disappeared, and returned on the fifth day of the month bleeding and perspiring. Shrine priests supposedly spread a rumor that the horses had taken part in a divine battle, of which the earthquake was a surface manifestation, and had returned victorious after defeating a Mongol army. Ryōi followed up on this rumor by ending volume 2 as follows: "At such times it is inevitable that there will be all sorts of rumors about things unseen and unheard. Those wise to the ways of the world will pay them no attention. Foolish women and children, however, shake with fear upon hearing such things, thinking that some big event has just taken place. Moreover, they become even more agitated upon hearing baseless popular explanations. Therefore, I conclude with these lines of verse: 'Danger, danger! I don't like thinking about earthquake rumors / They cause distress to high and low alike.'" [15] One of Ryōi's rhetorical strategies was to discuss a wide range of rumors, which readers probably would have heard, and then dismiss them.

Ryōi's skeptical attitude about rumors of supernatural events points to another possibility in reacting to earthquakes. Some early modern Japanese rejected or substantially minimized religious explanations of events in favor of what we might call rational or mechanistic explanations. A good example of an explicitly rational approach to earthquake-related phenomena is Jōtō Sanjin's *Account of Broken Windows* (*Yabure mado no ki*). He observed that the main buildings of many temples survived the shaking of the Ansei Edo earthquake with little or no damage. "Although people commonly attribute this result to supernatural intervention," Sanjin observed, he rejected such explanations in favor of rational principles (*kotowari*). He explained that the four eaves of the support beams in the temples "naturally served to balance the structures," so that despite the shaking they remained in equilibrium. For the same reason, Sanjin also rejected the explanation that intact bridges enjoyed divine protection.[16] In addition to advocating a rational approach to interpreting earthquakes, Sanjin helped sow the seeds of a claim that developed during the Meiji period and retains currency even today: traditional wooden structures, especially temples, embody a native wisdom for earthquake-resistant building techniques.[17]

There was no clear temporal dividing line, but by the late Tokugawa period, earthquake literature often contained skepticism about supernatural matters. Among the better-known works, the 1856 *Ansei Record* (*Ansei kenmonroku*) tends to treat natural forces in a mechanical, amoral manner. A tale of the death of a virtuous, filial daughter, for example, ends ambiguously. It points out that despite the adage that the workings of the heavenly way (*tentō*) reward goodness and visit calamities on what is evil, the daughter's case is an exception. The text speculates that what Buddhists call "residual karma" might be at work, but it takes no firm stance on this explanation.[18] *Ansei Record* also discussed the widely circulated rumor that strands of hair from the white horse of the Ise Shrine, found lodged in people's clothing, protected common people during the earthquake. The authors affirm this rumor, and add, "From the beginning our country has been a land of *kami* (*shinkoku*)." They suspend judgment, however, on whether or not what people thought was hair was indeed that.

Their analysis begins with the mass pilgrimages to Ise in 1830 and the material benefits pilgrims supposedly received, such as free food, drink, and transportation, with no difficulties for anyone, even women and children. Then the text questions whether such a utopia is even possible. It mentions the rumors of amulets that supposedly fell from the sky at that time, none of which remained extant after the event. Although skeptical overall, the author of *Ansei Record* does not deny absolutely the possibility that the Ise Shrine provided assistance in the present earthquake. After all, it is the oldest shrine in "Great Japan," going back thirty-seven hundred years.

The *Ansei Record* analysis does not stop there. It next mentions that in 1836, strands of hair were reported in Edo, which turned out to have been "malformed ki." On the other hand, examples from Chinese histories of strange things falling from the sky such as blood, crops, meat, or dirt might suggest that falling hair is within the realm of the possible. In the case of the

1836 strands of hair, however, somebody consulted Western science books and examined the material under a microscope. In this way, he confirmed that the material was not hair, and he printed his findings as a broadside. Sure enough, the *Ansei Record* author had a copy at hand and reproduced its content. The "hair" in 1836 turned out to be small worms produced by strange atmospheric conditions. The worms were dispersed by the wind and fed on plants, thus contributing to the crop failures of the time. In the end, *Ansei Record* defers the current question of falling hair to a later scholarly investigation. Skepticism regarding supernatural phenomena and a tendency not to speculate beyond verified facts is typical of *Ansei Record*.[19] It is interesting to note that falling white horsehair was mentioned in passing as early as 1596 in connection with the Keichō-Fushimi earthquake. Unfortunately, the temple document mentioning this phenomenon did not comment on its meaning or significance but simply listed the falling hair along with pebbles, sand, and dirt.[20]

Satire and parody were common manifestations of religious skepticism in connection with the Ansei Edo earthquake. A popular print, for example, featured a parody of the Immovable Wisdom King, Fudō Myōō, at a *kaichō,* the revelation of a hidden deity to the public for an admission fee. In it, a man surrounded by flames and looking much like a wealthy merchant appears standing on the head of a giant catfish. He is Hijō Myōō (Emergency Wisdom King), and his two attendants consist of the dark figure of a child's charred body on one side and the iconic Sensōji pagoda with its bent spire on the other. The sign announcing the *kaichō* reads, "Opening the doors of storehouses" (*Kura no kaihi*), and the text explains that the earthquake has forced the wealthy to open their storehouses in the form of charitable contributions. The print features many other layers of meaning and plays on words, but its overall point is that the cosmic forces caused a redistribution of wealth in society. Interestingly, the origin story for the deity is based on the Heian-period rebel Taira no Masakado ("Matakado" in the print), who was a popular hero figure in Edo at the time.[21] In this way, the print subtly acknowledges the potential for political upheaval in the earthquake, a potential that the redistribution of wealth effectively neutralized, as we will see in chapter 5.

A final point about religion is that a general tendency to suspect that earthquakes might be the result of divine punishment or warning was an ideal opportunity for certain kinds of social critics to amplify their messages. Conveniently, there was no need to posit any specific mechanism whereby a certain deity or group of deities caused or permitted the earth to shake owing to their displeasure. The simple fact that the earth had shaken destructively was sufficient to underscore critiques of society that ordinarily would receive little attention. The usual targets of such critiques were greed and luxury, not among rulers but among ordinary members of society. An early example comes from the Keichō-Fushimi earthquake. The Christian text *History of the Western Teaching in Japan* (*Nihon saikyōshi*) reported that some Japanese regarded warfare between the various kings of the Buddhist underworld or a battle between deities under the ground as having caused the earthquake. However, "in reality God [Tentei] has caused calamity throughout the country" as a manifestation of his anger at arrogance, licentiousness, and luxury.[22]

In a similar spirit, the Sōtō Zen priest and poet Ryōkan was walking from Wajima Village to Sanjō when the Sanjō earthquake occurred. He composed two poems based on his thoughts and observations. Part of "Jishingo no shi" (Post-earthquake poem) says,

> Reflecting on the past forty years
> The trend toward luxury in this world
> Has advanced like a galloping horse.
> All the more so, because for so long there has been peace and stable
>   government
> The feelings of the people have slackened
> . . .
> People think very highly of themselves and regard deceiving others as
>   great talent.
> Like piling up mud upon earth
> There is no end to their sordid deeds.[23]

Luxurious living and stable political conditions have warped people's sense of ethics and duty, and Ryōkan implies but does not explain a link between this state of moral degeneration and the earthquake.[24]

Perhaps the most important cultural product of the Sanjō earthquake was earthquake chants (*kudoki*). Starting in 1829, blind female musicians began to sing about the earthquake. A major theme was that those who have grown accustomed to material prosperity in a peaceful world and who are obsessed with a desire for private gain invite earthquakes:

Thinking about it
Society [*shi-nō-kō-shō*]
Has forgotten the way of
Confucius, the buddhas, and the deities
Led astray by personal desires
Regardless of high or low
Haughtiness in the extreme
Warriors abandon the arts of war
Lining up their
Pillows with merchants
Petty officials
Persecute those below
And wallow in luxury.[25]

After many verses explaining how bad things are, the final lines stress a positive note for the future, as if the earthquake has been both a price society has paid for its waywardness and an opportunity for its renewal. Although petty officials and warriors in general come under criticism in these chants, there is no specific critique of a domain government or the bakufu. From this point onward, moralistic chanted verse, both performed and written, spread throughout Japan in the wake of major disasters.[26]

The content of this verse and other attempts to link earthquakes with alleged moral failings of society ranged widely. The most prominent theme, however, was the corrupting influence of material desires, manifest in such forms as greed, luxurious living, deceit, or arrogance. In contrast with post-earthquake discourses of morality in Christian or Islamic contexts, briefly discussed in the introduction, issues connected with religious doctrine or sexual behavior were much less prominent in early modern Japan.

Earthquakes as Drama

Major earthquakes produced drama. The earth itself produced terrifying convulsions of destruction and created extreme conditions for some of the survivors. There was undoubtedly plenty of actual drama in the streets of the stricken cities and towns, but of course what remains for us are written records. Just as moralists and social critics used the occurrence of earthquakes to amplify their messages, writers seized on earthquakes as raw material for entertaining tales. Typically, such stories presented themselves as factual and accurate accounts, and some may have been. However, we have no way to verify the vast majority of earthquake tales. This point deserves highlighting, especially in light of the practice of some modern writers to take unverifiable early modern accounts at face value as evidence of possible coseismic signals.[27] Early modern earthquake tales are rarely useful as evidence of actual occurrences. Their value to scholars is in bringing social values into sharp relief.

Works such as *Ansei Record* were commercial ventures. Although much of the work clearly appeals to a visceral fascination with disaster and strange tales, the introduction stakes out high moral ground for the work: "Humans possess five states of ki and seven emotions. Amidst joy and anger, sorrow and elation, people's minds are apt to become disordered and they lose their ordinary presence of mind. If we deepen the scope of our contemplation during ordinary times, then even at times of extreme danger or ill fortune, we will be able to act without forgetting our social obligations and righteousness. Thus we present detailed exemplary tales that will inspire even ordinary women and children."[28]

Exciting drama was the selling point of *Ansei Record,* but much like Ryōi's *Foundation Stone, Ansei Record* postured as a beneficial work of moral edification. Elsewhere in the front matter, readers are told that the point of the work is not to inquire into the veracity of the stories it tells but to encourage loyalty, filial piety, and righteousness among our descendants.[29] On this point, *Ansei Record* contrasts rhetorically with *Ansei Chronicle,* which claimed to have carefully assessed the accuracy of the tales it reported.

Foundation Stone was the first work of commercial earthquake literature. It consists of three volumes. The first describes scenes of devastation in Kyoto, sometimes in graphic detail. These scenes include the mourning of parents whose children were crushed under a falling stone lantern, the collapse of part of the Gojō stone bridge, the collapse of the stone tower of the Kiyomizu Temple, the collapse of two hundred earthen storehouses around the city, the deafening noise made by the temple bells as they beat against their wooden ringers, and the affixing of sacred verses to gateposts in the midst of aftershocks.

The second volume deals with a variety of reports, tales, and rumors from areas beyond Kyoto. There were deadly landslides in Fushimi, for example, and several of the area's castles and the towns around them suffered serious damage. The Komatsu River in Kaga became obstructed and flooded. Even worse, there were rumors that Tsuruga in Echizen might be underwater owing to inundation from the ocean, and only four or five inns in Ōmi's Imazu remained standing. The landslides made for especially gripping narrative. Over one hundred died in Kutsuki, for example, and Katsuragawa suffered house-burying landslides from a partially collapsed mountain. Survivors above ground could hear the cries of buried victims. There were too few survivors, however, to mount an effective rescue, and after four or five days, the cries stopped. Such tales of death and destruction from the Kanbun earthquake were probably recycled during the 1830 Kyoto earthquake, as we will see. The third volume, discussed in the previous chapter, summarizes explanations of earthquakes derived from Chinese and Buddhist metaphysics, lists major past earthquakes, and ultimately seeks to reassure readers that society will flourish. *Foundation Stone* functioned as a template for later works that attempted to capture the total experience of major earthquakes.[30]

One of those works was *Record of Chastisement and Shaking (Chōshin hiroku)*, a product of the Sanjō earthquake. The bulk of the work consists of tales from the earthquake that illustrate moral principles. One section entitled "Greed is Difficult to Stop" starts with a story of a clothing store whose owner is so concerned with salvaging his money boxes that he stays too long in the burning structure, gets his foot caught under a beam, and has to amputate it with an ax. After much suffering, he dies from his wound. The moral of the story is stated baldly at the end: "Even tens of millions of coins" cannot purchase one's life. Following the story of the storeowner is a similar one about a merchant who burns to death after getting both legs caught in a beam. It is illustrated and shows a passerby burdened with an armload of moneyboxes unable to help.[31] The titles of other sections include "Honest to a Fault," "Things beyond One's Strength," "Chastity," "An Impressive Person," "Power of the Deities," "Loyalty and Courage," "A Heartwarming Person," "Cowardice," "Circumstances," and "Discussion." *Record of Chastisement and Shaking* contrasts with an anonymous collection of stories from the same earthquake of households that suffered death or injury, interspersed with lists of names of victims. These obituary-like tales report the details of the causes of death (usually falling beams) without drama or moralizing.[32]

After the Ansei Edo earthquake, shogunal bannerman Miyazaki Narumi was a relatively rare example of a warrior who wrote a detailed account of the event. Narumi tells of a guard at the Wadakura Gate whose arm became pinned between falling beams. Fires approached and, there being no choice, his comrades used a sword to amputate his arm. The guard died three days later, and "while his case deserves our pity," concluded Narumi, "compared with the extreme loyalty the old woman who sacrificed herself for her master displayed in the previous entry, the death of the guard desperate to save his own life seems pathetic." In the previous passage, Narumi had described an elderly servant who died to save a young daimyō wife.[33]

In another tale, Narumi reports at length on the travails of a man named Hayashida in Koishikawa. His pregnant wife and mother were sleeping on the second floor when the earthquake struck, the main beam pining both down. Hayashida called for a servant to help him, but the position of the two women was such that pulling up on the beam to relieve pressure on the mother increased the pressure on the wife and vice versa. The man decided to rescue his mother first, which crushed his wife to death. Narumi's evaluation of the matter was critical of Hayshida because he was in the prime of his life and should have thought the matter out more clearly. "While seeking single-mindedly to save his mother and casting aside his wife may resemble the filial way, a moment of thought followed by a call to neighbors for help would have permitted lifting both ends of the beam and saving mother, wife, and child."[34] Hayashida's lack of mental dexterity and the rigid either-or choice might have prompted critical readers to question the veracity of the tale. Indeed, many such tales from Ansei Edo or previous earthquakes featured moral choices cast in suspiciously stark and unrealistic terms.

Ansei Chronicle presented itself as on-the-spot reporting. Published without the proper authorization, it sold out its first print run in 1856 before being banned by a bakufu determined to reassert a measure of control over the publishing industry.[35] Ostensibly anonymous, as noted in chapter 2 the main author was almost certainly popular fiction writer Kanagaki Robun. Several different illustrators provided dramatic visual images to accompany Robun's prose. Some of the diverse material in the *Ansei Chronicle* was original, but much of it came directly or indirectly from other sources. For example, a script of dramatic chants about the earthquake called "Mikawa manzai" appears in *Fujiokaya Diary* (*Fujiokaya nikki*) amidst material relevant to the tenth month of 1855. Mikawa manzai was a form of popular drama with roots in the same geographic location as the Tokugawa house. The identical item later appeared in *Ansei Chronicle*.[36] Similarly, a tale about a man who thought his wife died in the earthquake and, upon seeing her later, fled thinking that she was a ghost appears first in Miyazaki Narumi's account and later in *Ansei Chronicle*.[37] The different Ansei Edo earthquake accounts picked up tales that were circulating in the weeks after the main shock, and it is usually impossible to know either the provenance or veracity of any particular story.

Comprehensive discussion of earthquake morality tales is beyond the scope of this study, but it is worthwhile to survey a few typical examples from *Ansei Chronicle*. The servant Kane was an exemplar of loyalty who gave no thought to her own injuries but instead rushed to the aid of the household head, trapped under a beam, and two other injured family members. Bakufu officials formally rewarded her with fifteen pieces of silver for selfless and devoted attention to duty. The next story tells of a selfish, disloyal manservant for explicit contrast with Kane. The obvious conclusion was that the two "are as different as black and white."[38]

In another tale, the courtesan Mayuzumi, whose childhood name was Kane, sold her combs and hairpins to purchase over twelve hundred cooking pots for the temporary shelters the bakufu had installed. She received an award from the bakufu and became famous. It turned out that her motive in part was to meet the parents she had not seen since age seven.[39]

One tale designed to invoke a sense of pity or compassion was that of the tragic case of Miuraya, a Shin-Yoshiwara brothel whose women all took shelter from the shaking in an underground storehouse but perished in the fires that swept through the district.[40] The psychological impact of the destruction of the elite brothel district was profound, and it was frequently the setting for tragic earthquake tales.

For dramatic impact, there was the story of samurai Yamaguchi Shūhei, whose arm was pinned beneath a beam. As fire approached, he told his son to cut off his arm with a sword, thus resulting in a filial piety–related drama. At first, the son could not bear to do violence to his father, but he finally succeeded. Father and son escaped the flames, and their lord later rewarded the son.[41] Contrast the ending of this tale with Miyazaki Narumi's comments on the Wadakura Gate guard in a similar situation.

As we have seen, a tale found in both Narumi's account and *Ansei Chronicle* featured a wife (named Yasu in the *Ansei Chronicle* version) who found herself trapped under a beam. As the fire drew nearer, she implored her husband to save himself for the sake of their three children. He vowed to raise the children and there was a dramatic, emotional farewell before he made his escape. As the flames consumed the beam, however, it became light enough to permit the wife to escape. When she went to rejoin her husband, he saw her approaching, disheveled and covered with soot, and assumed that she was a ghost.[42]

In another tale, firefighters heroically risked their lives to save dozens of women of a prominent household in Ogasawara. The firefighters prospered from generous monetary rewards, and *Ansei Chronicle* concluded from this tale that one's prosperity or decline is a matter of one's own actions, which are repaid in kind.[43]

The tales mentioned thus far are examples of extraordinary deeds that fall, at least barely, within the realm of what was possible. Although *Ansei Chronicle* portrayed its accounts of these matters as simply reporting the facts on the ground, it is hard to imagine there was not some degree of editorial work to make the tales fit perceived reader interest and stock emotional and moral points.[44]

Other *Ansei Chronicle* tales deal with supernatural or uncanny phenomena. For example, there are two tales of supernatural warnings. In one, a fox possessed a man to warn him that there would be an earthquake on the second day of the tenth month. He left town and returned to do business after the shaking had subsided, at which time the fox dropped its spell over him.[45] In another, a woman saw her deceased father in a dream, and he ordered her to flee from the house in which she was serving as an apprentice. The earthquake struck the next day.[46]

Other tales involve the power of the deities or tales of karmic retribution. In one, the wooden statue of Kannon at Sensōji's Thunder Gate (Kaminari Mon) disappeared at the time of the earthquake. The rumor was that it knew the earthquake was about to strike and left the area. The temple denied the veracity of the rumor, saying that a priest had removed the statue for repairs. However, the rumor of Kannon's predictive powers persisted and expanded. A statue of a wooden horse behind the main hall of the temple was found with mud on its feet after the earthquake. The conclusion was that Kannon mounted the horse to escape the shaking.[47] In another tale, a cruel-hearted couple favored their biological daughter and neglected their adopted but very filial daughter. They even tried to sell her, but they perished in the earthquake and she survived. This tale was but a slight variant on a common folk motif, one of the more obviously nonoriginal parts of *Ansei Chronicle*.[48]

The specific values advanced in early modern earthquake morality tales are the usual fare of sincerity, loyalty, filial piety, courage, generosity, and so forth. More significant is the overall message of individual agency. By remaining calm, thoughtful, and resolute in a crisis, people can do much to extricate themselves and those around them from danger. This attitude is similar to that underlying popular religious devotion. In this realm, too, people can act to enhance their chances of survival or prosperity by forging alliances with deities through devotion. Indeed, especially late in the Tokugawa period, we find some Japanese rejecting the power of deities entirely in favor of the power of human intelligence. By 1856, some writers even imagined technology based on this intelligence that might predict, prevent, or mitigate the harmful effects of earthquakes.

Earthquakes and Media Sensationalism

The Kyoto earthquake of 1830 was a major event. However, given a death toll of two to three hundred, reports such as the following seem sensationally exaggerated: "At the fourth hour in the afternoon [the day after the main shock] another great earthquake occurred, and Kyoto is turned upside down. Even if the (second) earthquake had not occurred, the whole city would have burned down from the fires that were raging."[49] Rumors spread rapidly, including reports of landslides and devastation in places far from the city that could not have experienced significant shaking. Indeed, in Wakasa, some eighteen villages were rumored to have sunk into a sea of mud caused by a (nonexistent) tsunami. In his detailed study of the Kyoto earthquake, Miki Haruo speculates that the rumors of sinking villages came directly out of literature published 130–170 years earlier about severe floods and other disasters in that area, including the Kanbun earthquake.[50] Writers in 1830 might even have mined much earlier sources from the region. For example, one account from the 1596 Keichō-Fushimi earthquake described a tsunami nearly wiping out the coastal villages in the Hyōgo area.[51] Miki's conclusion is that the mass media was responsible for rumormongering by exaggerating the destruction in 1830.[52] Sensational exaggeration was evident two years earlier in connection with the Sanjō earthquake, but circumstances such as its relatively remote location minimized the broader psychological impact of that event.[53]

Kitahara Itoko has followed up on Miki's basic conclusion. One factor she mentions was that the mass media network of the day uncritically recirculated material of dubious validity. Takizawa Bakin, for example, collected earthquake reports that included many inaccurate tabloid prints. Works written in jest such as *Strange Earthquake Tales of the Capital*

Manzairaku Dance (Jishin kidan miyako manzairaku) and *Strange Winds Letter* (*Fūkaijō*) by "Catfish" were absorbed into other works. The former, for example, became part of *Kasshi Night Tales* (*Kasshi yawa*), and the latter became part of *State of the Floating World* (*Ukiyō no arisama*). *State of the Floating World* includes three other tabloid accounts. Reacting to this situation in popular media, five days after the main shock local authorities in Kyoto ineffectively banned the sale of any material promoting rumors.[54]

Kitahara mentions that a small, pamphletlike record of past earthquakes (*jishin nenpyō*) circulated after the earthquake. At its end, the document's author or authors state the motivation for its creation and publication: "This book was not written to alarm society. A variety of rumors about the Kyoto earthquake are floating around regions far and wide, and we have heard that people with relatives or acquaintances in Kyoto are worried. Therefore, we have written a summary of earthquakes and hope that it will calm the anxieties of people in distant places."[55] This passage is nearly identical to the final page of *Record of Earthquakes in Japan* (*Honchō jishinki*) discussed in the previous chapter.[56] Most likely the work to which Kitahara refers is *Record of Earthquakes in Japan* in content if not in title. This statement of intent points to the phenomenon of broadside prints circulating widely, spreading exaggerated news of Kyoto's alleged destruction to other parts of Japan. Moreover, it is probable that the lurid accounts of the Kyoto earthquake served in part as models for press coverage of the Ansei Edo earthquake.[57] In any case, from the time of the Kyoto earthquake onward, media coverage extending throughout Japan became a feature of major earthquakes and other calamities.

Other factors combined with the competition to sell broadside newspapers contributed to the exaggerated sense of doom in Kyoto and beyond. Not only was Kyoto the imperial capital, but 1830 was an *okage* year. During the Tokugawa period, outbreaks of spontaneous mass pilgrimages, called *okage-mairi* or *nuke-mairi*, periodically occurred on a multiprovince, regional scale. People would drop what they were doing and make their way to the Ise Shrine, often with a bacchanalian atmosphere prevailing among the group. The pilgrims would march and dance their way toward Ise, living off donations. There were three especially large-scale occurrences in 1705, 1771, and 1830, with as much as a third of the population of some areas participating. *State of the Floating World* conjoins the 1830 earthquake and the pilgrimages. According to its account, the deities created the earthquake to punish the people of Kyoto for alleged stinginess in assisting the pilgrims.

The people of Yamato and Kawachi, on the other hand, were generous in helping Ise Shrine visitors, and the deities rewarded them with a bountiful harvest.[58] Again we see the common rhetorical device of using earthquakes to amplify an author's social message.

The Kyoto earthquake made an impression throughout Japan much greater than its estimated 6.5 magnitude and relatively modest death toll might suggest. It also occurred just before what would turn out to be a poor harvest and the start of the Tenpō Famine. From the vantage point of hindsight, many accounts of Japan's history regard the Bakumatsu era as starting in 1830. Even from the perspective of the 1840s or 1850s, it might reasonably appear that 1830 was the start of a series of major changes and upheavals. I take up this and related matters in chapter 5.

Earthquakes as Creators

Although seemingly counterintuitive at first glance, earthquakes were not solely destructive phenomena. In many cases, their shaking created opportunities for economic profits, opportunities for governments to demonstrate their benevolence, and even significant new agricultural real estate. The rebuilding phase of urban earthquakes usually increased the demand for labor, thus raising wages for a wide range of commoner occupations. Unskilled laborers could find abundant work hauling debris, and skilled workers in the construction trades could command many times their normal wages. Although terrifying, earthquakes could be a short-term gold mine for urban survivors. An entry in *Tadokoro's Record* (*Tadokoro-shi kiroku*), an account of local conditions near the southern tip of the Kii Peninsula in present-day Wakayama Prefecture, reports that the wages of carpenters and day laborers are high because of the recent Hōei earthquake and tsunami.[59] The Hōei earthquake was not the first time that laborers enjoyed high pay during the rebuilding phase after earthquakes. Nevertheless, the idea that earthquakes can profit certain segments of society appeared with increasing frequently in earthquake-related documents and literature over the course of the eighteenth century. In the immediate aftermath of the Ansei Edo earthquake, even before rebuilding had begun, catfish prints included graphic elements that anticipated windfall profits for those in the construction trades.[60] A similar attitude appears in the following lines from a series of short earthquake-related verses in the aftermath of the 1830 Kyoto earthquake:

Face lights up with joy . . . Carpenters
Truly enlivened . . . Plasterers[61]

Simple economics may have been one reason early modern earthquakes in Japan never produced large-scale civil unrest. The likelihood of windfall profits for many survivors was a strong incentive for people to work within the existing social system.

Starting in the eighteenth century, there was a steady growth in the extent to which governments responded to disasters by providing relief to the victims of floods, volcanic eruptions, fires, and earthquakes. In describing state responses to the 1891 Nōbi earthquake, Gregory Clancey points out that the emperor assumed a high profile in the relief effort. Clancey further suggests that this response was a radical departure from the past: "Here was a much different Japanese government than the one that had lived in mysterious seclusion in Kyoto or Edo. . . . Here was also a different moral universe than the one that surrounded the Ansei quake in 1855. In the Nōbi earthquake, the previously sharp Japanese sense that disaster critiqued the order of things and redistributed good and ill fortune was buried behind a façade of a collective 'human emotion' and concerted 'national' action."[62]

Certainly there were differences between 1855 and 1891 in the nature of the relief effort, and the shogun and emperor were indeed secluded figures in 1855 and earlier. Key elements of the "moral universe," however, were substantially the same. Indeed, those who saw disasters as a critique of society could be more focused in their blame during the modern era. This phenomenon was especially apparent after the Great Kantō Earthquake of 1923, when diverse segments of society regarded the disaster "as a warning, an act of divine punishment, or a necessary evil."[63] In any case, it was the duty of the state to provide relief in the wake of disasters, whether in 1855 or 1891. One reason the emperor was so prominent in 1891 is that he was by then the unquestioned center of the sole *kuni* ("country," a term that in early modern times usually referred to provinces and large daimyō domains) in Japan. During the Tokugawa period, the state was much less centralized, and domain governments closest to the afflicted areas often provided much of the relief.

Documents from the Hōei earthquake in *Tadokoro's Record* describe grants of relief rice from the local lord. For example, seven villages received an initial grant of over eleven *koku* of relief rice and small quantities of cash for the purchase of agricultural tools. The same villages later received over 37 *koku* of rice, 124 *koku* of barley, over 28 *koku* of rice bran, and 45 bags of salt.[64] Another account from the same region, near present-day Tanabe City, includes the details of relief rice distribution approximately ten days after the earthquake from the local government office. This office also distributed basic supplies for constructing temporary shelters.[65] From approximately the time of the Hōei earthquake, indications of official relief frequently appeared in earthquake documents. By the end of the century, charitable relief from relevant governments was standard practice and had developed into a predictable pattern that included providing food and shelter for the poorest members of society for one to two months.[66]

After the Sanjō earthquake, domain governments in the region provided relief. One account near present-day Shibata City, Niigata Prefecture, describes the damage and adds that the district government office decisively took action, distributing five hundred bales of rice.[67] Another document describes the efforts of temples to provide temporary shelter for the homeless.[68] The bakufu also sent aid to the region, and local emergency granaries assisted in the effort.[69]

Domain governments likewise intervened effectively after the Zenkōji earthquake struck. Collapsing sides of mountains and other landslides created natural dams in fifty-one places in the Matsushiro domain and forty-one places in the Matsumoto domain.[70] The collapse of Mt. Iwakura created the most important of these dams, consisting of mud and sand approximately one hundred meters high. It blocked the Sai River near its headwaters and buried the two villages of Iwakura and Magose in the process. As the days went by, more villages became inundated by an expanding lake that reached forty kilometers in length, eventually swallowing up thirty villages.[71] The potential for a much greater disaster was present owing to the likelihood that the dam would give way and inundate areas downstream with powerful floodwaters. To deal with the danger, the Matsushiro domain mobilized thousands of retainers. They began work on an embankment to help channel the water. The domain sent messengers to downstream areas urging villagers to flee to higher ground. The messengers also visited mountainous areas, ordering villagers to assist any strangers who might show up. Domain officials established a watch over the dammed area on a nearby mountain and floated small, unoccupied boats on the growing lake to serve as an early indication of the dam starting to give way. When the dam showed evidence of collapsing, those on duty were to light a signal fire.[72]

After half a month, the dam showed no signs of collapse, and those who had fled downstream villages were approaching physical and psychological exhaustion. Some even returned to their villages and began working the fields. On the evening of May 27, the dam broke with a great roar, and twenty days' accumulation of water burst onto the Zenkōji Plain. At Koichi, a town along the river at the edge of the plain, the water level briefly reached twenty meters. The wild rush of water lasted about four hours, washing away over eight hundred dwellings and inundating over two thousand more with sand and mud. Only about one hundred people perished, however, thanks to the domain's emergency measures.[73] The official response to the disaster encompassed much more than encouraging flood safety measures. In some areas, the earthquake wiped out sources of food, drinking water, and shelter all at once. Domain relief efforts included establishing temporary shelters, providing basic food and supplies, cash grants, and emergency loans for rebuilding. The Matsushiro domain, for example, provided some 1,268,000 meals as of the ninth day of the seventh lunar month, and it had by then distributed 13,429 *ryō* in cash.[74] Eight years later, the bakufu's emergency relief agency, the Machigaisho, undertook many of these same steps in dealing with the aftermath of the Ansei Edo earthquake. In many instances, early modern earthquakes functioned not so much as divine chastisements of rulers but as opportunities for governments to demonstrate their benevolence. Of course, when earthquakes caused beneficial changes, rulers were probably pleased by suggestions that their virtue was somehow connected with the process.

For all of the destruction it caused, the Kanbun earthquake brought some dramatic benefits to the coastal area near the Japan Sea. The earthquake caused as much as 4.5 meters of uplift between the shore of Wakasa Bay and the eastern part of the Mikata-gokō Lake, which created new agricultural fields for the Obama domain in Wakasa and in adjacent areas.[75] Indeed, these new fields significantly increased the productivity of that domain, allowing it to recover from the effects of the Kan'ei Famine of the 1640s.[76] According to one local account discussing hydraulic geography, Mikata Lake tilted to the west and areas that had once been under water in the eastern sides of three lakes became dry land. It concluded by declaring, "The virtue of our ruler became the profit [*rieki*] of the land."[77] Presumably, this virtuous ruler was the lord of the Obama domain. Similarly, according to *Origins of Tidal Flat Kanzeon* (*Higata Kanzeon engi*), the appearance of over three thousand *koku* of new fields was the result of the deep desire of the domain lord to save the people.[78] Although such praise was little more than a figure of speech, in these examples we can glimpse a limited invocation of the classic view that the state of the ruler's virtue can influence powerful cosmic forces.

The Genroku earthquake lifted the southern part of the Bōsō Peninsula by four to five meters, creating new land.[79] Indeed, the name of the Nojima Promontory derives from its having been an island prior to the earthquake. The visible part of the one-time island is now a hill, and the ring of land surrounding it is called the Genroku Terrace. Fishing settlements quickly developed on the flat ground of the terrace, and the JR Tateyama Station is located atop the terrace today.[80]

Taking examples from present-day Chiba Prefecture, Tateyama experienced one meter of uplift, and land around the village of Iida rose by 5.5 meters. The village of Numa received 720 meters in extra shoreline, and nearby Kashiwazaki Inlet expanded its shoreline by about 200 meters.[81] In many cases, uplift also extended the tidal flats in coastal areas. Sometimes, local villages were able to develop this land into sites for houses or, with a source of fresh water nearby, into agricultural fields. Boundary lines between villages sometimes had to be redrawn, irrigation arrangements rearranged, and, of course, the new fields had to be sold or distributed. In general, uplift in this earthquake added to the prosperity of many coastal areas, but the new shape of the landscape also caused disputes over resources.[82]

About a century later, in 1804 in Kisakata (Akita Prefecture), the situation was even more dramatic. As the sun rose on July 10, local residents were amazed to see islands rising from a swampy lagoon. An M7.1 earthquake created the "Ninety-Nine Islands of Kisakata." The earthquake thrust upward the wet fields famously described twenty-five years earlier by Matsuo Bashō. Today's wet fields in the area were once the ocean floor, and today's hills were islands prior to the earthquake.[83] Similarly, a broadside from 1847 reported on the creation of an earthquake mountain in a farmer's field in Tango, named "Prosperity Mountain" (Hōeizan).[84] The 1854 Tōkai and Nankai earthquakes also produced significant uplift. In the Muroto Peninsula in Kōchi, for example, uplift raised the port of Murozu about 1.2 meters, making it difficult for large ships to dock there.[85] Around the western edge of today's Tōkaidō Shinkansen Fukikawa Railroad Bridge, uplift of between one and three meters created "Kamahara Earthquake Mountain" and additional agricultural fields for Kamahara. A local song inspired by this gift of land from the cosmic forces included the line

"Earthquake, oh earthquake, please come back—once for my generation and twice or three times for my descendants."[86]

Being located at the intersection of several plate boundaries, Japan's islands have been shifted by earthquakes in vivid fashion from time to time. Several earthquakes during the Tokugawa period caused significant uplift, creating new land. Although the claim of *Ways of Earthquakes in Japan* (Honchō jishin no shidai) that earthquakes created Mt. Fuji and Lake Biwa overstated the power of seismic activity, earthquakes did contribute to the contours of Japan physically, socially, and in the realm of the imagination.

Japan

An account of the Genroku earthquake described it as a "deadly disaster unprecedented since the creation of Japan."[87] Similarly, *Night Tales of Kasshi* contains a brief essay on the Sanjō earthquake, which argues that 1828 was an unlucky year in general because of flood damage in several areas of Honshu and typhoons in Kyushu so powerful that Dutch sailors reported they had never experienced such storms. Indeed, the overall situation was "unprecedented since ancient times."[88] It was the nature of severe earthquakes to stimulate large-scale thought with respect to time, place, and intensity. Those who lived through early modern earthquakes and wrote about them repeatedly described the events as "unprecedented" (*zendai-mimon* or *mizō*). One reason was the timing. Major earthquakes were frequent, but not so frequent as to be in the living memory of most Japanese at any given time. Someone who experienced a severe earthquake was usually doing so for the first time in his or her life. Therefore, the event would have seemed unprecedented. Moreover, when writers qualified their use of "unprecedented," for greater impact they often employed reference points extending far into the past and widely across a geographic zone that included all of Japan.

Characterizing an earthquake as unprecedented also raised the possibility that something had gone terribly wrong with the world. Indeed, one possibility for interpreting major earthquakes was to regard them as vectors of cosmically ordained change, typically marking the end of an era. In traditional Inca society, for example, major natural disasters periodically occurred regardless of human merit, turned the world upside down, and announced an apocalypse that began a new era. Similar views of disasters can be found among Aztecs, Hopi, and Ute tribes in North America.[89] Classical Chinese notions of the mandate of heaven adhered to a similar logic. Despite the practice of changing era names after some large earthquakes or other disasters, the approach to earthquakes in Japan tended to be quite the opposite: an affirmation of the unchanging, unshakable foundations of society, often embodied in the imperial court. As we will see in chapter 5, however, one line of interpretation of the Ansei Edo earthquake did come close to suggesting that the event was a harbinger of momentous change. In this sense, the Ansei Edo earthquake was a partial exception to the propensity of Japanese commentators to normalize earthquakes.

In response to initial characterizations of an earthquake as unprecedented, a second wave of commentators typically pointed out that the recent earthquake was not an event without precedent. A typical example is *Account of Chastisement and Shaking (Chōshin hiroku)* from the Sanjō earthquake, the final section of which begins with a discussion of past earthquakes, starting from the time of Emperor Tenmu. It explains that although earthquakes seem to be unprecedented, looking into the past reveals that they occur regularly.[90] This historicizing approach was part of a rhetoric of reassurance. At the core of this rhetoric was a claim that Japan is resilient to seismic upheavals, not to mention lesser calamities. Earthquakes also served as opportunities to discuss and characterize Japan in other ways.

Rhetoric of Reassurance

The obvious aspect of the rhetoric of reassurance was its historicizing approach, which transformed earthquakes from anomalies to at least semiregular, seminormal events. One facet of this approach was a tendency for some earthquakes to prompt a discovery of specific past earthquakes as being especially relevant or significant in hindsight. Sometimes the reason was simply to remind readers that a massive seismic event had taken place in the same area far in the past. The 1707 Hōei earthquake, for example, promoted discussion of the Hakuhō earthquake of 684. *Kokuryōki*, an account of Tosa after the Hōei earthquake, contains a description of the Hakuhō earthquake, the earliest recorded massively destructive earthquake in Japan. Hakuhō submerged approximately twelve square kilometers of farmland in Tosa. "From Hakuhō to Hōei 4," the discussion concludes, "is a period of 1,022 years"[91]—long but not unprecedented. Indeed, the term "1,000-year earthquake" is now common in Japan as a result of the 2011 Great East Japan Earthquake.

Sometimes the previous major earthquake to shake the same area served as a beginning point in discussing a particularly significant sequence of earthquakes from the perspective of the present event. Materials from the Ansei Edo earthquake, for example, frequently regarded the Kyoto earthquake and especially the Zenkōji earthquake as significant recently past earthquakes. In a typical example, the catfish print *Earthquake Taiheiki* (*Jishin Taiheiki*) features a large group of catfish clad in robes bowing down to the Kashima deity, who hovers above them. The catfish represent past earthquakes, and the lengthy text features each of them apologizing for their disruptive deeds. The series of earthquakes begins with the 1703 Genroku earthquake, the previous major earthquake to shake Edo, and includes the Hōei earthquake, the 1804 Dewa earthquake, the Sanjō earthquake, the Kyoto earthquake, the Zenkōji earthquake, and the Odawara earthquake.[92] It may be significant that very few works from Ansei Edo included the recent Ansei Tōkai and Nankai earthquakes in their configurations of important earthquakes from the past. Perhaps these large, destructive earthquakes occurring within the same era were too close for comfort.

In modern times, this historicizing approach has transformed into the idea of a natural, cyclical occurrence of catastrophes. In Japan and elsewhere, the idea of periodically occurring "characteristic earthquakes" (*koyū jishin*) is deeply entrenched in both general and scientific thinking, despite recent criticism from prominent seismologists.[93] More generally, there is a tendency to speak of major fires, floods, storms, and other catastrophes as "___ year" events, revealing an assumption of at least semiregular occurrence. As anthropologist Susanna M. Hoffman explains, discussing contemporary U.S. society, "Earthquakes and avalanches are not only due, but should they miss their deadline, 'overdue.' The informal cyclical thinking is perpetuated by scientists and their ominous yet vague forecasts." Moreover, through this informal notion of cyclical occurrence, "Americans, like other peoples, culturally manipulate catastrophes so that they appear, both in prediction and certainly in aftermath, to be anticipated and normal."[94] The historicizing of earthquakes in early modern Japanese rhetoric was an early form of this normalizing strategy.

Another approach was to extend the geographic scope of the discussion to include China and possibly other countries. After the Kanbun earthquake, for example, *True Record of Gen'en* (*Gen'en jitsuroku*) reflected on past earthquakes. It begins with a question posed to shogunal scholar Hayashi Shunsai (Gahō). After a summary of the death and destruction caused by the current earthquake, Shunsai is asked whether such events have taken place in the past. He responds, "Such things have occurred in both Japan and China." His first example is a 1293 earthquake during the reign of Emperor Fushimi that caused death and destruction in Kamakura, which he pairs with a Chinese earthquake "also at about that time" (1290, according to the sexagenary cycle date) that killed seven thousand. Next is a Japanese earthquake from 1290, followed by a Chinese earthquake that killed 22,300. The passage ends by stating that there were many more earthquakes besides these.[95] We have seen other examples such as *Foundation Stone* that discussed Chinese earthquakes, often stressing the occurrence of such events during the reigns of sagely kings and emperors. One obvious effect of such discussion was to point out that earthquakes have occurred during well-governed reigns and on a worldwide scope. More subtly, at least vis-à-vis earthquakes, Japan and China were roughly on a par: two lands of great antiquity, well governed, occasionally tested by shaking, and resilient. Indeed, we will see in chapter 5 that it was common in both medieval and early modern rhetoric to declare Japan to be "small" or otherwise seemingly insignificant on the world stage as a setup for asserting its strength and greatness.

Comparisons with China helped situate Japan in a broader world. Similarly, early modern earthquakes had the effect of broadening perspectives beyond the local level. At a time when the primary geographic identification of most Japanese was with their *kuni* (province or daimyō domain) or even a smaller community, powerful seismic waves served as reminders of the broader realm of Japan. Because most large earthquakes were indeed felt over a wide area, it was relatively easy to imagine them shaking a place called Japan. One account of the Hōei earthquake in Tosa claims that "no parts of Japan were spared from this earthquake, though Kyoto suffered only a little. In general, the area along the Tōkaidō suffered the most destruction. The routes in Kyushu encountered some damage and Shikoku was especially hard hit. Within Shikoku, Tosa suffered great damage."[96] Although ultimately focused on Tosa, the shaking of the earthquake brought forth an image of Japan and some of its major regions. The writer, describing the degrees of damage here and there, does not even mention vast areas of Japan, yet he imagines the earthquake to have shaken the whole country.

Following an earthquake of a similar scope and intensity, the Ansei Nankai earthquake, a writer in Osaka, after describing some of the far-flung areas affected, concluded, "Truly, this earthquake seems to be an event for the entire country of Japan."[97] Earthquakes often perceptually refreshed the geographic links between one's local region and the larger entity of

Japan. One commentator on the Zenkōji earthquake, for example, wrote, "As if attached to a flying horse, Shinano [Nagano] is the high ground of the imperial country [kōkoku]. Therefore, it is the source of many rivers of various provinces [shokoku]."[98] The focus in most pieces of earthquake literature tended to be local, but the seismic event often situated one's home province in the larger entity of Japan.

Resilient Land of Deities

Situating one's local, badly shaken area within the larger framework of Japan was part of the process of reassurance. A town or city might lie in ruins from shaking imagined to have been felt everywhere in Japan, yet the "land of deities" has a long history of resilience in the face of such events. We have already seen several examples of verse stating in various iterations that Japan's long reign of sovereigns would continue regardless of how much the earth might shake. Typically, that sovereign was the emperor, even though the emperor in early modern Japan was more a concept than a specific person. The following short poem from the Kyoto earthquake is a typical example:

> The coolness we faced as the shaking made us prostrate ourselves
> to the outside
> Was a sign that our venerable reign [miyo] continues.[99]

Earthquakes focused on Edo, however, might stimulate similar talk of the shogun's resiliency. Discussing the Genroku earthquake, Nectar Chronicle (Kanrosō) states, "Because the shogun [taikun] is militarily strong and deeply benevolent, the population, in the manner of children, will come at once to repair the damage, thoroughly excelling in construction. Well governed and suffering not from hunger and cold—can we not be grateful for the pride of our indomitable spirit? Someone in the palace composed the following lines of verse about the present earthquake: 'Grasping the state for a thousand generations, some shaking / has tried to dislodge our unmovable, august reign.'"[100] Even in this case, however, the shogun's role in partaking of or nurturing the people's "indomitable spirit" was rooted in the longer, deeper tradition of the imperial court. By contrast, during the next big shakeup in 1855, we will see that the shogun did not appear so invincible. He and his city required active intervention and assistance from the principal deity of the imperial court.

Earthquakes highlighted Japan's long history and deep roots in the imperial court. They also foregrounded another key characteristic of Japan. It was a land of deities, a *shinkoku*, presided over by thousands of *kami*, buddhas, and bodhisattvas. These deities helped ensure the country's prosperity, especially in the form of bountiful harvests. *Earthquake Tales from the Capital (Jishin kidan miyako manzairaku,* 1830) points out that "Japan is a land of deities unchanged for 10,000 generations." It mentions "Amaterasu's Shrine," the Kashima Shrine, and includes a variation of the verse we saw in *Foundation Stone:* "No matter how much the earth might shake, the Foundation Stone will not be dislodged, as long as the Kashima deity is present."[101] *Earthquake Tales from the Capital* was written for popular entertainment, which is precisely why this religious vision of Japan in connection with the earthquake is especially worthy of note. Of course, in the aftermath of an earthquake, feelings toward the deities, who were far from infallible, might be mixed. One account of the Zenkōji earthquake speculates that the event might have been a cruel joke by "mad *kami,*" directed not only at people from "our country" (Shinano) but also at those from "other countries"—that is, other provinces. Such a situation was especially cruel because the people dwelling in "our divine country" devoted themselves day and night to the deities of heaven and earth.[102]

Writing the in the month after the Zenkōji earthquake, a local resident named Miyasaka began an essay by stating, "Our country is a land of deities, superior to all others and whose people possess superior wisdom. The five grains flourish, and we receive divine favor." Even such a magnificent country as Japan, however, cannot escape natural disasters. Miyasaka explained that in times of calm, people inevitably incline toward luxury, thus angering Heaven, which sends down calamities such as landslides, floods, and earthquakes. He discusses several ancient examples of earthquakes before turning to the Sanjō earthquake, "in recent times." The discussion of past earthquakes then narrows in geographic scope to "our Shinano country," going back to an earthquake in 887, during the reign of Emperor Kōkō. Mention of recent strange weather conditions and "people out of touch with the deities" sets the stage for a discussion of the present conditions in the aftermath of the Zenkōji earthquake.[103] Miyasaka's account, of course, includes the strategy we have already examined of using the fact of a recent earthquake to amplify a social critique.

In Miyasaka's account, "Heaven" seems to be a higher power than "the deities," which likely reflects a variety of late medieval and early modern religious thinking known as "*tentō* thought," in which *tentō*—literally "Way of Heaven"—takes on some of the qualities of a supreme deity. In any case, what is especially noteworthy is Miyasaka's assertion that Japan's characteristic of being a land of deities makes it superior to other countries. This point may seem obvious, but the history of the term *shinkoku* is complex. In many medieval conceptions of *shinkoku,* for example, Japan's being a land of *kami* reflected its remote, inferior status vis-à-vis the sagely land of China and the Buddha land of India.

I examine the topic of *shinkoku* in more detail in connection the Ansei Edo earthquake in chapter 5. From what we have seen thus far, however, it should be clear that in the context of earthquakes, Japan was as much or more a religious entity as a geopolitical one. The frequency with which writers used the term *shinkoku* increased after earthquakes, reflecting the claim of resilience. Because Japan enjoyed the protection of thousands of deities, even a severe earthquake would not destroy its social fabric. Indeed, an examination of the relative frequency with which the term *shinkoku* appeared in written discourse throughout Japan's history reveals that it serves as a rough barometer of social anxiety. The term first appeared in the context of the possibility of a joint Tang-Silla invasion of the Japanese islands in the seventh century. Later, it appeared with great frequency in connection with the armed disputes between temples, shrines, and the court during the era of cloistered emperors, in establishment attacks on new forms of Buddhism in the early Kamakura period, and in connection with the Mongol incursions of the thirteenth century.[104] As we will see, the term *shinkoku* appeared frequently in popular discourse in 1853 and 1854, in connection with Perry's visits. Not surprisingly, major earthquakes evoked levels of anxiety roughly on a par with foreign military incursions or large-scale political unrest. As a reaction to this anxiety, writers often pointed out that Japan was a land full of deities with a long history. Moreover, despite individual flaws and limitations, as a whole these deities were benevolent. They bestowed benefits on Japan and its people, especially bountiful harvests. Surely, a country thus blessed would rebound from any recent earthquake.

Imagined Connections with Recent Events

Finally, current and recent events often intersected with earthquakes and conceptions of Japan to produce webs of associations. In the background, of course, was a general notion that earthquakes and other major events are not random occurrences. Moreover, people everywhere tend to confuse correlation and causality, early modern Japan being no exception. The 1854 Iga-Ueno earthquake occurred several months after the signing of the March 1854 Treaty of Kanagawa, which formally initiated diplomatic ties between Japan and the United States. The treaty also served as a template for establishing relations with several other foreign countries. Diaries and other documents from the time sometimes link the two events in the sense that they were both newsworthy and took place in sufficiently close proximity to imply some kind of causal connection. For example, Perry's visits, their effect on commodity prices, the establishment of diplomatic facilities at Shimoda, and recent visits of foreign ships near Osaka contextualize the Iga-Ueno earthquake in *Osaka Earthquake Record* (*Ōsaka jishinki*).[105]

As was the case with Iga-Ueno, several commentators linked Perry's black ships with the Ansei Tōkai and Nankai earthquakes. Both events were newsworthy, unsettling, and close together in time, and suggested that major changes were taking place. One letter writer spoke of military mobilization in connection with the black ships and the earthquake and concluded that "people's minds are deeply agitated," a point he reinforced in a classical Chinese poem.[106] Another writer mentioned the appearance of foreign ships in Shimoda and, according to rumor, in and around Osaka. He added that rumors are circulating and that the upheavals of the earth and sea have caused people's minds to worry.[107] A diary entry from Kyushu describes the Ansei Tōkai/Nankai earthquake as "a great earthquake that shook the entire country of Japan: Osaka, Kyoto, and Edo." It then explains that this earthquake occurred at about the time that four American ships entered Uraga. The passage continues, vaguely linking the arrival of the ships with "many" deaths in the area. Moreover, the author spins a wide web of associations that includes the arrival of two foreign ships in Nagasaki. At the end, he summarizes the following coincidental (or not) occurrences: the great earthquake, the situation in Edo, the deaths of twenty-four thousand people, the loss of a large ship (probably the Russian vessel *Diana*), collapsed houses, and much else. The situation is a "great anomaly," unprecedented even in connection with earthquakes of the past.[108] The association of Commodore Perry, earthquakes, and other recent events continued for the next year and a half, reaching a peak during the Ansei Edo earthquake. We will see that in many minds, the seismic events of recent years and the coming of exotic foreigners became elements in a larger process of social change and upheaval.

Conclusions

Earthquakes helped shape the social and imaginative contours of Japan. They encouraged reflection on core social values and individual virtue and the broader world of deities and cosmic forces. Earthquakes encouraged a broad view of geography that situated local communities within larger imagined communities. Survivors of an earthquake could and did imagine that throughout wide areas, perhaps throughout Japan as a whole, others had experienced a similar terrifying event. Mass media encouraged such views, even while often exaggerating the extent of earthquake damage, especially from 1830 onward. Earthquakes literally created more of Japan, and may have created certain of its iconographic features in the imaginations of some writers.

More important, earthquakes set up a cycle of social discourse. It was difficult to regard major earthquakes as random events. Moreover, most Japanese would have experienced a major earthquake only once in their lifetimes. Such a momentous event would have seemed like an unprecedented upheaval of society, thus exacerbating anxiety levels among survivors. These anxiety levels produced rhetoric assuring survivors that earthquakes (1) are common historical experiences in both Japan and China; (2) can be explained in rational terms (see chapter 2); and (3) cannot harm Japan because it is a resilient country protected by benevolent deities. Those deities might not be able to prevent earthquakes, but they were capable of sustaining society in the aftermath. Furthermore, as time went on during the Tokugawa period, governments played an increasingly active role in providing relief in the wake of natural disasters.

We turn next to a close examination of what is arguably the most important earthquake in early modern Japan, the Ansei Edo earthquake. This earthquake was in many respects a culmination of cultural trends up to that point. It also cast a long shadow over the Meiji era and beyond to the present day.

CHAPTER 4

The Ansei Edo Earthquake

One of the many accounts of the Ansei Edo earthquake was *Record of the Great Earthquake and Great Storm* (*Ōjishin ōkaze kenmonki*). Its title refers to the typhoon that struck Edo approximately ten months after the earthquake, so this account reflects approximately a year of hindsight. The "General Notes" section starts with a concise summary of the earthquake: "First, there was the great Kantō [Genroku] earthquake on the twenty-second day, eleventh month of Genroku 16, with a great fire on the twenty-ninth day. Now, because it has been 154 years between then and Ansei 2, the passing of years lulled people into thinking that Edo was a city without major earthquakes. Therefore, the population was unprepared for a great earthquake. Thinking that Edo was prone only to fires, people built plaster-walled dwellings and erected plaster-walled enclosures with tile roofing. Such construction became the major cause of injuries." Moreover, the destruction wrought by the earthquake was uneven: "In this situation, the visible effects of the earthquake were severe in some places and weak in others. Thatched-roof structures generally suffered little damage, whereas the walls of earthen storehouses collapsed right and left, causing countless deaths from debris striking the men and women who emerged to flee. The collapse of compound walls of mansions, stone walls, objects thrown off from such structures, falling stone lanterns, and so forth harmed many people."[1] One important aspect of the Ansei Edo earthquake was that its impact varied widely as a function of topography, underlying geology, construction materials, and social geography.

The Ansei Edo earthquake defies simple characterization. This chapter presents a description of the patterns of destruction, benefit, relief, and rebuilding that occurred in connection with this event. In this context,

I argue that the earthquake was both a destructive and a creative event whose effects differed owing to the interplay of local geology and social geography. Moreover, conclusions about this earthquake in the limited English-language literature, especially with respect to patterns of destruction, are incomplete and misleading. Rumors and exaggerations about the earthquake were common both during its immediate aftermath and throughout the next century. An understanding of what actually took place on the second day of the tenth lunar month, 1855, and in the months thereafter, is a prerequisite for understanding the earthquake's broader significance, which is discussed in subsequent chapters.

Patterns of Death and Destruction

The overall death toll, civilian and military, amounted to about 1 percent of Edo's population. Fires were a problem for the first two to three days, usually in areas experiencing the most severe shaking. Winds were calm, however, and most deaths occurred because of falling objects and crushing forces. This situation contrasts sharply with the Great Kantō Earthquake of 1923, in which fire was the main cause of death. In 1855, several factors promoted perceptions of exaggerated death tolls, and rumors circulated of one hundred thousand or more killed.

Aftershocks kept the residents of Edo on edge. Astute observer Mishima Masayuki recorded aftershocks in detail in *After the Shaking* (*Nai no nochimigusa*). He recorded at least one aftershock for almost every day of the tenth month. Aftershocks continued into the first few days of the eleventh month, but by then their frequency had greatly diminished.[2] The unease prompted by these aftershocks caused many townspeople initially to abandon their houses and dwell in hastily erected huts by the roadside. "People erected various huts and for a while endured rain and mist." Moreover, "There were also many residents of areas near the shore on land reclaimed from the sea in Fukagawa who fled because of fears that a tsunami might arrive," a point made in several accounts.[3] These temporary dwellings had become a problem in some areas to the extent that three days after the earthquake, the City Magistrate's Office (Machibugyō, hereafter "City Magistrate") declared, "There are so many temporary huts erected by those fearing aftershocks that military personnel cannot pass through. City officials must rectify this situation."[4] *Ansei Record* (*Ansei kenmonroku*) compared the 1830 Kyoto earthquake to the current event: "The current earthquake has a lower frequency [of aftershocks] than Kyoto did, and their intensity is not as severe. Similar to the situation reported in *Thoughts on Earthquakes* [*Jishinkō*], however, many people became frightened and slept in the main streets. Therefore, an anonymous person posted a notice at Nihonbashi that read, 'There is no reason [*ri*] that another earthquake will occur, so set your minds at ease and return home. If not, the night air will eventually make you ill.'"[5] Accumulated knowledge from recent past earthquakes undoubtedly shortened the time required for many of Edo's residents to realize that the worst was behind them and to shift into rebuilding and recovery mode.

A Soil Base Disaster

In the Ansei Edo earthquake, the most important causal factor in determining the severity of ground motion, and therefore destruction, was the soil base. Ground motion was least severe in upland areas such as the Yamanote Highlands. It was more severe but still relatively moderate in lowland areas of the city built on solid ground. Such areas typically experienced a level of five using the JMA seismic intensity scale (*shindo*). Ground motion, destruction, fires, and loss of life were most severe (level six) in areas built on unconsolidated fill. In other words, former wetland areas filled in with dirt or debris after Edo became Japan's de facto capital suffered dramatically more destruction than other parts of the city.

Owing to a phenomenon called material amplification, ground motion increases in severity as the base soil becomes softer. Alluvial soil amplifies seismic waves compared with sedimentary rock, and silt or unconsolidated fill produces the greatest degree of amplification.[6] An increase in wave amplitude, a vertical shift in the path of seismic waves, and basement rock trapping seismic waves are the reasons for this amplification.[7] Liquefaction is also common in areas built on unconsolidated fill. The 1995 Hanshin-Awaji (Kobe) earthquake was so destructive "because the strongest shaking was squeezed into a 3km wide plain between mountains and Osaka Bay" and many structures were built "on harbor fill."[8] Moreover, in the 1923 Great Kantō Earthquake, "disturbances in alluvial material were particularly marked" to such a degree that in one area, "potatoes were extruded onto the ground" from the shaking of soft soil.[9] The basic geophysical principle that ground motion is most severe with soft, unconsolidated soil is what made the Ansei Edo earthquake primarily a "soil base disaster" (*jiban saigai*), to use Kitahara Itoko's term.[10]

One of the few studies of the Ansei Edo earthquake in English is Andrew Markus' fine essay, "Gesaku Authors and the Ansei Earthquake of 1855." A work based in large part on a close and insightful reading of primary sources, Markus' essay has been influential as a reference point for comparisons of the Ansei Edo earthquake with later seismic disasters. Unfortunately, Markus' description of the damage, while not incorrect as far as it goes, is misleading:

> Whether because of greater proximity to the epicenter of the earth-quake, or softer alluvial soil, or whether because of a higher residential density, poorer quality of building construction, and a lack of open spaces for refuge, the plebian areas closest to the Sumida River were the most severely affected. Destruction was greatest in the Shitamachi (downtown) districts of Honjo, Fukagawa, Kameido, Asakusa. In contrast, Yamanote (uptown) districts like Yotsuya, Akasaka, and Ichigaya, topographically as well as economically superior, were far less susceptible to the worst effects of the catastrophe. While the Ansei earthquake deeply affected all sectors of the population and all facets of urban life, its 'social bias' against the least affluent classes of citizens is noteworthy.[11]

Markus does indeed identify all the significant factors (aside from random chance) that determined degrees of destruction, but without pri-oritizing them. Moreover, in an attempt to demonstrate socio-cosmic bias against the "least affluent classes," Markus leaves out a major component of the picture, namely the devastation of Daimyo Lane (Daimyō Kōji) and some other elite warrior neighborhoods. The reality on the ground did not conform so neatly to socioeconomic levels.

Markus' characterization of the overall damage pattern has influenced other studies. In his book on the Nōbi earthquake, for example, Gregory Clancey drew comparisons with 1855. Relying substantially on Markus, Clancey described the damage pattern as "disproportionately concentrated in the residential districts of the lower town. The great temples and shrines, and the residences of the Shogun, daimyō, and samurai had been com-paratively less affected than those of the artisan and merchant classes. . . . Tokugawa elites had also commandeered relatively high, rocky ground for themselves away from the alluvial plains where the effects of earthquakes were most severe, but where common people often had no choice but to cluster and build."[12] In effect, Clancey logically extends Markus' description of the overall damage pattern to claim that elites suffered significantly less than commoners did. Moreover, this claim became part of Clancey's assessment of the 1891 Nōbi Earthquake: "That the physical infrastructure of the state or other elites (such as the new industrial concerns) should be particularly susceptible to damage from natural disaster was arguably unusual in the Japanese experience, and a reversal of the age-old order in which suffering had been more clearly demarcated along lines of wealth and power."[13] The notion that elites fared comparatively well in the Ansei Edo earthquake is in part a function of sources and data. We have very accurate data on commoner death and destruction owing to systematic surveys orchestrated by the City Magistrate and extensive essays, diaries, and other written accounts. Warriors wrote a few earthquake accounts, but most accounts were the products of townspeople. Deaths in military households were approximately as numerous as those in civilian households, but the military data is much more dispersed and difficult to interpret.[14]

Another reason for a tendency to focus on the earthquake as a disaster mainly affecting townspeople of modest means was the general Meiji-era debates over architecture that Clancey analyzes in detail. Advocates of native, wood-based construction, as opposed to foreign, brick-and-mortar construction, tended to stress that well-built wooden structures have historically performed well in earthquakes. Because 1855 was the most recent test case, they emphasized that only shoddy structures in poor neighborhoods suffered severe damage. For example, in the context of debates about the best type of construction for the imperial palace, government architect and wood advocate Tachikawa Tomokata argued that in 1855, "only the low degree houses' had fallen, while those 'above the middling level' sustained only the slightest damage."[15] This Meiji-era concern with construction quality reinforced a view of the Ansei Edo earthquake as an event whose destruction neatly divided along socioeconomic lines. It also contributed to the notion that traditional, premodern construction techniques evolved in response to earthquakes and were somehow more resistant to violent shaking than are modern structures. This claim is generally inaccurate. Indeed, today's scholars need to make adjustments when using damage to structures in estimating the level of historical earthquakes on the JMA seismic intensity scale (whose current iteration dates from 1995), because across all categories of construction, premodern structures were weaker than those of contemporary or recent vintage.[16]

As Clancey and Markus point out, earthquake damage indeed differed significantly along a divide of highlands versus lowlands. This high
versus low topographical division, however, did not closely correspond to
economic categories or the formal status categories of warrior versus commoner. Most commoners did live in lowland areas, but not all commoner
neighborhoods were located on a poor soil base. Warriors occupied both
high and low areas. The mansions of Matsudaira Buzen-no-kami and Hotta
Bitchū-no-kami, for example, were built in Kanda's Ogawamachi on unconsolidated fill atop what was once the Ogawa Marsh. It might well have been
the worst place in Edo to erect a building, and these two daimyō did not
even attempt to rebuild in that location after the earthquake obliterated
their residences.[17]

The most prestigious district in Edo was the area within the outer castle precincts (o-kuruwa uchi), corresponding to present-day Marunouchi
business district and parts of Ōtemachi. Most of this area was located atop
a poor soil base. Indeed, much of it had been marsh or open water, the
Hibiya Cove, prior to 1608. Classified as "foundational" (kaname), the area
was the location of elite fudai daimyō residences and major bakufu offices
such as the Hyōjōsho (the highest judicial organ), one of two Kanjōsho
(treasury offices; the other was within the castle), and both City Magistrate
offices. The area was the heart of Tokugawa authority outside of Edo Castle
itself, and it experienced severe, dramatic damage. According to Noguchi
Takehiko, it was as if the earthquake "volitionally" targeted the center of
bakufu power.[18] The death rate in Daimyo Lane was unusually high (736),
and Nihonbashi neighborhood head Saitō Gesshin provided a vivid account
of the fate of this area: "The mansions of the domain lords came crashing
down all at once. Right away, fires broke out here and there, and the sound
as large timbers burnt and tiles collapsed resounded through heaven and
earth—a repeat of the roar of the earthquake."[19] The passage continues with
an extensive listing of the destroyed mansions.

Much of this elite zone faced a commoner district located just across
the outer moat of Edo Castle and separated by the bridges Gofukubashi and
Kajibashi. In terms of today's landmarks, the commoner district ran along
present-day Chūō-dōri between Nihonbashi and Kyōbashi. This district
suffered modest damage, mostly to storehouses. By contrast, just across the
moat, the warrior side was a mass of collapsed buildings, fires, and general confusion. In 1600, only yesterday in geologic time, this devastated
zone was Hibiya Cove. The nearby commoner neighborhood was located

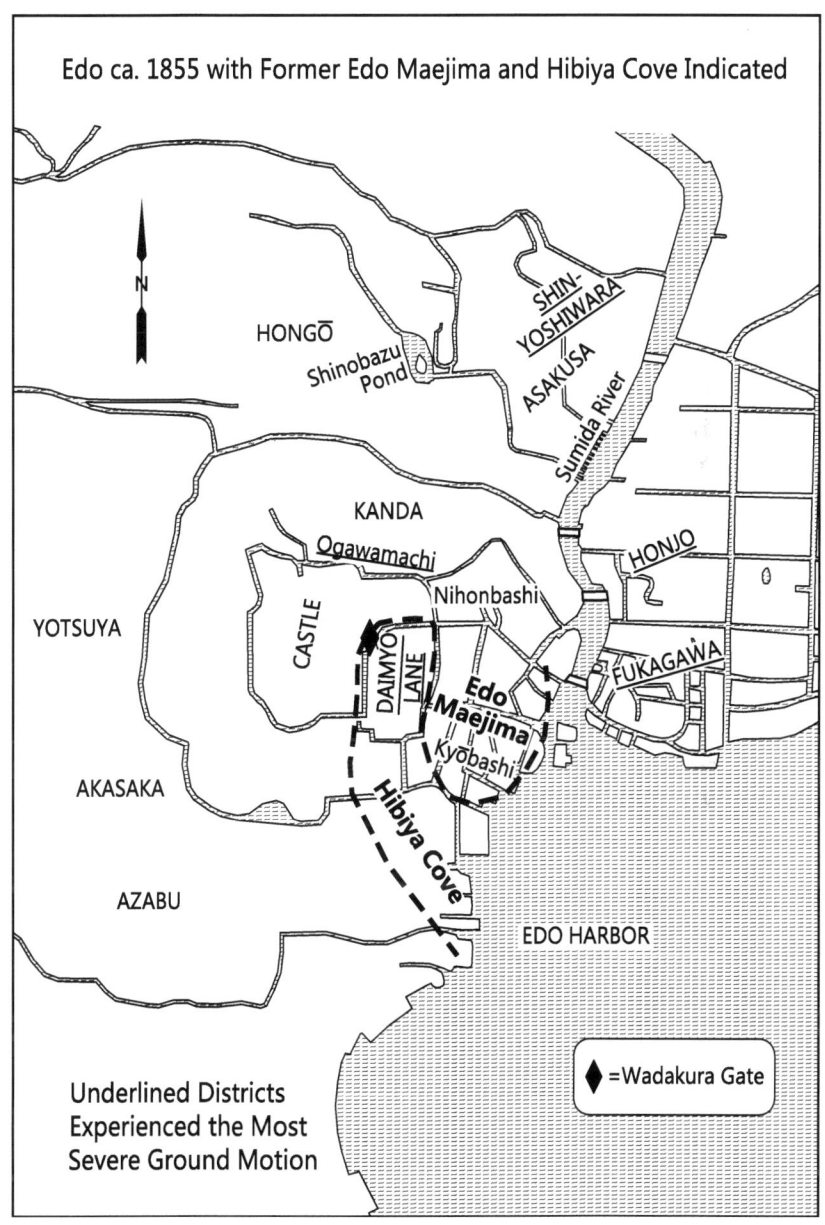

MAP 2 Produced by Jeffrey Smits, Cherokee Drafting Specialists.

on solid ground—what in 1600 was an outcropping of land known as Edo Maejima, a wave-eroded plateau. The inner moat of Edo Castle marked the sixteenth-century boundary between land and sea. The Wadakura Gate, for example, was once the site of an underground storehouse (*kura*) that extended under Hibiya Cove. The Aizu mansion was located just to the east of this gate, atop the former cove, and the earthquake destroyed it. In terms of today's landmarks, Tokyo Station is on solid ground, and to the east of it is the former Edo Maejima. Yūrakuchō Station and Shinbashi Station are atop the former Hibiya Cove (map 2).[20]

The landscape of Edo underwent a major transformation during the first decade of the Tokugawa bakufu. Beginning in 1603, a small army of workers from several domains began transporting dirt from nearby Kandayama to fill in Hibiya Cove.[21] The newly created land became prime real estate. It was as if the 1855 earthquake shook the boundaries of an older, forgotten Edo back into sight, tracing the contours of pre-bakufu times via dramatically differing patterns of destruction.[22] Jōtō Sanjin, the pen name of Iwamoto Sashichi, described leaving his residence and crossing Gofuku Bridge to enter Daimyo Lane and then the Wadakura Gate. After explaining that this area shook especially hard, he characterized it repeatedly as "fire earth" (*kachi*), possibly a term he created (it appears in the *Classic of Changes* in connection with the thirty-fifth hexagram).[23] Fires in the Ansei Edo earthquake tended to correspond closely to the areas of most severe shaking, and according to a young police official at the scene, "Daimyo Lane burned intensely everywhere."[24] One additional factor that made the fires in Daimyo Lane especially dramatic was that several daimyō were in charge of the recently constructed offshore artillery batteries. Their mansions contained stores of potassium nitrate (saltpeter), which ignited in the flames.[25]

Sanjin's account of the damage summarized the situation by explaining that upland areas shook mildly, whereas lowland areas (including Fukagawa, Honjo, and the castle precincts) shook severely. "This can be called a principle of nature," he concluded.[26] Other earthquake accounts made a similar point. *Record of the Times* (*Jifūroku*), for example, after describing the devastation in several low-lying areas, says, "Indeed, highland areas shook lightly and damage to structures as well as deaths were few."[27] Nearly every comprehensive account of the damage described the extreme devastation in both commoner and elite areas built on soft ground. In most instances, the authors of these accounts regarded elevation as the key variable, not soil base. Although upland neighborhoods were all built on a firm base, the soil base of lowland areas varied greatly, as we have seen.

At least one writer, however, precisely identified the relevant variables owing to his knowledge of Edo's history. After a detailed accounting of damage in the downtown areas, Mishima Masayuki explains, "Since the building of Edo Castle, this was an unprecedented shaking. Starting with the *gosanke,* and extending to daimyō, bannermen, *gokenin,* retainers, peasants, townspeople, and everyone else, very few were able to escape this disaster."[28] He also speculates that daimyō, bannermen, bakufu retainers, and other warriors would likely seek to conceal casualties in their households. "Based on what I witnessed," Masayuki concludes, "amidst the collapses of residences in Daimyo Lane, there must have been many cases of injury and sudden death." He continues his analysis, explaining that most daimyō household members live in two-story, long houses with limited exits, especially for the women of the household. Overall, Masayuki estimates casualties in military households as twice those of the townspeople.[29] With greater precision than Sanjin, Masayuki explains the exact reason why the earthquake devastated areas such as Fukagawa, Honjo, and Daimyo Lane so severely: "Places built on land that long ago was created from the sea shore naturally shake more severely in an earthquake."[30]

Although the quality of the soil base was the most important variable in determining rates of death and destruction for everyone, population density and quality of construction were significant secondary factors. Arguably, the worst location for someone when the shaking started would have been the upper story of a Shin-Yoshiwara brothel. Also dangerous would have been a rented back-alley shop in the Tomigaoka Hachiman area of Fukagawa. Residents of the same neighborhood in comparatively spacious shops facing a main street were more likely to have survived.[31] Sakuma Chōkei was a nineteen-year-old police official on duty when the earthquake struck. He described conditions in the back alleys in part as follows: "In the neighborhoods, those who lived in alleyway shops panicked to escape through the narrow paths. Their steps broke the drain-covering boards, causing sprained ankles or the inability to move. Many children were trampled to death. There were numerous instances of injuries from falling roof tiles."[32] Undoubtedly, the people who endured such harrowing conditions and survived would have been inclined to imagine levels of death and destruction in Edo much higher than the statistics collected at the time indicated.

These statistics came from two surveys ordered by the City Magistrate and carried out by the neighborhood heads. The City Magistrate was able to act quickly after the earthquake struck. The shaking destroyed or damaged nearly every mansion in Daimyo Lane and major bakufu offices such as the Hyōjōsho and the lower Kanjōsho. Almost inexplicably, however, offices connected with city governance and relief survived with relatively little harm. After a lengthy listing of damage to elite mansions, Saitō Gesshin pointed out that many of the military households in the Marunouchi area were making do in makeshift, temporary structures. In a separate entry, he next pointed out that the northern and southern City Magistrate offices were unharmed except for damage to long houses attached to the northern Machibugyō. Moreover, Gesshin explained that the three offices of the Machidoshiyori (city administrators below the City Magistrate and above the neighborhood heads) were undamaged. Especially significant, the office of the Machigaisho (sometimes spelled Machikaisho), Edo's emergency relief agency, also escaped serious damage. Its grain storehouses were badly damaged, but they did not catch fire, and the grain was available for relief efforts.[33] It was as if the cosmic forces had struck the center of bakufu power but spared precisely those parts of the government in a position to provide relief to the townspeople.

Surveying the Damage

Knowing the precise extent of deaths, injuries, and property damage was essential for an effective relief effort. Two days after the main shock, the City Magistrate offices ordered all neighborhood heads to conduct a survey of deaths, injuries, and several types of property damage. The next day, officials extended the order to cover damage to structures.[34] The neighborhood heads conducted their surveys at that time and conducted a second round of surveys during the middle of the same month. There were practical problems in compiling accurate statistics. For example, most of the fires had been extinguished or brought under control by the end of the fourth day. The neighborhood heads did not include the charred remains of burned buildings in their figures for collapsed buildings, thus undercounting that category. Furthermore, some neighborhood heads counted buildings (*ken*) while others counted roof beams (*mune*), even though in some areas a single beam supported the roof of as many as six residential units. Despite such inconsistencies, the survey results enabled bakufu relief officials to direct resources to the neediest areas. Moreover, statistics for commoner casualties derived from the surveys were generally accurate, and it is likely that city officials leaked the casualty data to the popular press to counter wildly exaggerated rumors about the death toll.[35]

Actual townspeople deaths ranged from zero in the Azabu area to 1,186 in Fukagawa. Comparing adjacent areas, the relatively more densely populated commoner district around Nihonbashi, Nakabashi, and Kyōbashi suffered sixty-nine deaths, compared with over seven hundred estimated deaths just across the moat in less densely populated Daimyo Lane. Again, soil base trumped all other factors in determining casualty rates. Several contemporary sources mention the survey results, usually divided into twenty-one districts (*kumi*), with male and female death and injury statistics for each. Saitō Gesshin lists a grand total of 4,293 deaths and 2,759 injuries, and *Fujiokaya Diary* (*Fujiokaya nikki*) is nearly the same, listing 4,303 deaths and 2,784 injuries.[36] Mishima Masayuki provided a more detailed breakdown for deaths. He recorded commoner deaths (no injuries, no male-female breakdown) for each of roughly four hundred neighborhoods and arrived at a figure of 4,616.[37] Jōtō Sanjin's *Account of Broken Windows* (*Yabure mado no ki*) claimed a figure of 3,895 commoner deaths based on reports submitted to the City Magistrate offices, with very little breakdown of the figures. Later, he surmised the total military, civilian, and shrine and temple death toll as "surely 10,000."[38] *Record of the Times* also states a commoner death toll of 3,895, based on figures from eighteen districts (not twenty-one), which is surely why the total is slightly less than what Gesshin or Masayuki reported.[39] Assuming a slight undercounting owing to unofficial residents, townspeople who may have died while serving in warrior households, or simple oversight, total civilian deaths were in the range of four to five thousand. Given an 1855 commoner population of roughly 540,000, the overall civilian death rate was less than 1 percent.[40] Even if deaths in military households were somewhat higher than those of townspeople, the Edo-wide death toll would still have been roughly 1 percent. The Ansei Edo earthquake was deadly, of course, but the overall fatality rate was low—a point that some bakufu officials probably wanted the public to know.

In the confusion and panic of the main shock and its aftermath, however, rumors abounded of much higher death rates. *Land and Sea Earthquake Record* (*Kainai jishinroku*) reported three thousand deaths in

Shin-Yoshiwara and eighty thousand in the city as a whole.[41] *Record of the Great Edo Earthquake of Ansei 2* (*Ansei itsubō Edo daijishin hikki*) gives a figure of one hundred thousand deaths for the whole city.[42] After providing a detailed assessment of death and damage leading to a grand total estimate of ten thousand deaths, Sanjin states that he is aware of rumors of death tolls of "more than 30,000, more than 50,000, and in extreme cases more than 210,000 people, all found as hearsay in written materials." He assures his readers that "they are all rumors not to be taken seriously" and worries that if transmitted to later generations, such exaggerations will be taken as facts.[43] Indeed, we see in chapter 6 that claims for this earthquake ranging from several tens of thousands of deaths to one hundred thousand were common during the Meiji period and into the twentieth century.

The confusion following the main shock, combined with the tendency of rumors to multiply bad news, in part explains the exaggerated casualty statistics. Another contributing factor was that high-profile areas of the city such as Daimyo Lane and Shin-Yoshiwara experienced particularly dramatic devastation. More concretely, the death toll was high enough, particularly in the worst-hit areas, that the supply of coffins, or even reasonable substitutes, was inadequate. One of several iconographic images of the earthquake was people carrying bodies to temples in a wide array of makeshift containers. A large visual image in the sensational booklet *A Close Look at the Earthquake and Fires* (*Jishin narabini shukka saikenki*) shows a street full of people, most of whom are transporting bodies. Barrels, stretchers, large boxes, carts, liquor vats, stretchers, boards suspended on ropes, and possibly even a chest of drawers serve as coffin substitutes.[44] *Ansei Chronicle* (*Ansei kenmonshi*) includes a similar image.[45] After claiming that the day after the earthquake "the dead formed a mountain," *Record of the Great Edo Earthquake of Ansei 2* explains that tubs, sake barrels, noodle boxes, and charcoal bags served as containers for the dead, some of whom were even wrapped in ropes and carried on poles.[46] *Record of the Times* mentions tubs, urns, sake barrels, oil barrels, sugar barrels, *sōmen* barrels, and even neighborhood water cisterns.[47] Miyazaki Narumi in *Accounts of the Ansei Earthquake* (*Ansei itsubō jishin kibun*) also mentions several of these containers and points out that soy sauce barrels served as coffins for children.[48] The sight and smell of so many bodies, some of which were charred or otherwise disfigured, must have been a source of social tension until the excess bodies were finally brought to temples after several days.[49]

Significance of the Offshore Batteries

Several other characteristics of the damage likely influenced popular perceptions of the earthquake. One was the destruction of all five offshore artillery batteries. Commodore Perry's first visit to Japan was a relative sideshow for the people of Edo because most of the action took place near Uraga. Following Perry's initial visit, bakufu officials assembled plans to construct eleven offshore artillery batteries (o-daiba) on artificial islands to guard the approach to Edo Castle, but a lack of funds scaled the project back to five.[50] When construction began in late 1853, ordinary townspeople began to realize the potential seriousness of Perry's activities.[51] Construction of the batteries required the mobilization of a large workforce to create artificial islands by piling up sacks of dirt. The daily wage was 250 mon of copper cash, and the project resulted in the issuing of a new silver coin, the Kaei Isshugin, known popularly as o-daiba gin, to expedite wage payments. A popular verse went, "For the sake of our lord (kimi) we build an island on spring nights / And receive profits for ourselves." The verse was a parody of a classical poem, and it is an indication of the significant impact that the batteries made on the local population, both economically and psychologically.[52]

The artillery batteries were a visible symbol of bakufu military power, and trusted daimyō allies were in charge of staffing and operating each of them. Beneath the surface of the bottom of Tokyo Bay is a large valley or trough created by the flow of the Old Tokyo River roughly twenty thousand years ago. Sediments have gradually accumulated over this ancient valley, and the artillery batteries were constructed on this poor soil base in water about five meters deep.[53] Of course, such details were unknown even to the most perceptive of the commentators of the day. The collapse of all five batteries, therefore, appeared to the public as a direct strike by the cosmic forces on bakufu military capabilities.

The number two battery, manned by soldiers from the Aizu domain, figured most prominently in earthquake accounts because of the false rumor that its magazine had exploded and killed everyone.[54] Indeed, maps of the city in popular publications depicting earthquake damage sometimes showed the five artillery batteries, but with number two as a burned-out ruin.[55] When the earthquake struck, the soldiers on duty had no idea what had hit them, and some thought they had come under attack from an American vessel. Fear that indeed the magazine would explode hampered rescue efforts, and in the end, fourteen Aizu soldiers died.[56] The earthquake rendered the shogun's capital defenseless against naval bombardment.[57] The wide attention the batteries received helped further erode the image of the bakufu as a powerful military organization, a topic discussed in chapter 5.

Miyazaki Narumi briefly discussed the batteries and the earthquake in the context of geomancy. He explained that some people thought Edo was, or should be, immune to large earthquakes, and that the Genroku earthquake was comparatively mild. In their view, construction of the artillery batteries constricted the flow of the tides and blocked an "earth vein." As the flow of ki built up in the vein, it suddenly burst, causing the earthquake. In this view, if the artillery platforms were rebuilt, Edo would periodically suffer severe earthquakes. Pointing out that inland areas such as Kyoto, Shinano (Zenkōji), and Sagami (Odawara) had recently experienced earthquakes, Narumi dismissed the theory as "the words of a madman."[58] There is no indication from other sources that many people took this idea seriously. It is an example of the variety of theories that circulated in the wake of the earthquake, as well as the dominance of the general idea that obstructions in flows of energy under the ground can cause earthquakes.

Significance of Shin-Yoshiwara

Another aspect of the damage with a strong psychological impact was the devastation of the elite brothel district, Shin-Yoshiwara (often "Yoshiwara" in documents). As Markus explains, "The earthquake struck as activity in the district was at its height, its alleys most crowded with lights and pleasure seekers. As buildings crumpled and fires erupted simultaneously in all five divisions of the quarter, a throng of prostitutes, entertainers, clients, and sightseers stampeded toward the single main exit, the O-mon, 'Great Gate.' However, the gate was twisted and would not open; efforts to lower the emergency 'gangplanks' over the moat that ringed the quarter were unavailing. Of all the scenes of surprising horror that night, the New Yoshiwara [Shin-Yoshiwara] was the most vivid in the collective consciousness."[59]

Nearly every earthquake account dwells on the devastation of Shin-Yoshiwara. According to *After the Shaking*, for example, "Yoshiwara is widely known to have been horribly charred. All that is left of it is two or three earthen storehouses."[60] *Record of the Times* explains that "particularly in Yoshiwara," because the earthquake struck at the peak of activity, "thousands of men and women in the midst of reveling in the upper and lower levels were instantly shaken down to the ground, soon to be consumed in fire. Survivors were rare."[61] Saitō Gesshin explains that the earthquake "leveled Yoshiwara all at once," and that the dead were "too numerous to count." In a subsequent passage, he points out that an investigation on the fifth day (the survey ordered by the City Magistrate) put the death toll at 630, and that adding in visitors and others from outside areas, the grand total would probably amount to one thousand or slightly less.[62] *Fujiokaya Diary* describes the "heavenly world" of Yoshiwara as turning to hell in an instant on the night of the second day.[63] Miyazaki Narumi pointed out that the stench from the burning objects and bodies was difficult to endure. He also told a strange tale of the charred remains of a Western-style military drill team found among the Shin-Yoshiwara carnage, whose weapons included thirty Gewehr guns and seven Japanese-made firearms. An investigation determined that this drill team, one of several in Edo at the time, had been showing off its skills for the amusement of the courtesans that night. None of its members survived.[64]

On the fourth day of the eleventh month, the City Magistrate dealt with the destruction of Shin-Yoshiwara by issuing a directive permitting the establishment of temporary brothels (*karitaku*) in twenty-four locations for five hundred days.[65] Following the destruction of Shin-Yoshiwara in 1812 by fire, the bakufu adopted a similar policy.[66] Word of this development spread quickly. Several earthquake accounts mention the temporary brothels, listing their locations in some cases. The brothels were also a prominent theme in catfish prints.[67] Customers frequenting such establishments soon after the main shock were not elite merchants but townspeople in the construction trades or other lines of work that enjoyed windfall profits. The earthquake created a temporary transfer of wealth from elite merchants to ordinary townspeople and often subsequently to the temporary brothels.[68] The devastation of the outer precincts of Edo Castle and the offshore artillery batteries appeared as a cosmic assault on the bakufu. Similarly, the charred ruins of the elite brothel district appeared as an assault on the playground of Edo's wealthy men. The temporary replacing of wealthy merchants by townspeople who in ordinary circumstances could never afford to visit the licensed quarters was, in the eyes of these townspeople, the quintessential meaning of world renewal.

Storehouses

The Ansei Edo earthquake had an affinity for smashing storehouses. Even areas of Edo that suffered little or no damage to other structures usually experienced collapsed storehouses. Nearly every account of the earthquake discusses storehouses, and collapsed or badly damaged storehouses litter the landscape in many visual images, whether in books or broadsides. *Record of Strange Earthquake Tales* (*Jishin kidanroku*) begins with a tale in which the collapsing storehouse of a pawnbroker's residence killed eighteen people.[69] According to *After the Shaking*, large houses with plaster walls suffered great damage in the earthquake. Owing to fire danger, storehouses also feature plaster construction, "but in an earthquake they become exceedingly dangerous."[70] Heavy, plaster-walled structures like storehouses were especially susceptible to the short-period seismic waves associated with many of Japan's inland earthquakes (*chokkakei jishin*). Moreover, in 1855 nearly all storehouses were of heavy construction because in 1842 the City Magistrate issued orders to the neighborhood heads that in the interest of fire safety, henceforth all storehouses must be constructed with mud plaster walls and tiled roofs. The roofs on all house extensions and outbuildings were also to be tiled. The collapse of such structures caused approximately half the earthquake injuries and many deaths.[71]

The earthquake caused architectural change as the city rebuilt, as Miyazaki Narumi explains: "In the present earthquake, the majority of tile roofs collapsed. Therefore, those dwellings whose roofs remain are discarding the tiles. Tiles set on top of a thin wood lattice [*doibuki*] are lined up in a manner called '*ichimatsu*,' which works for wind, rain, and earthquakes. Deeming two-story structures unsuitable, people are removing the top story to make the structure flat and then roofing with thin boards [*kokera-buki*]. Roofers are very busy these days and their wages keep rising more and more."[72] Elsewhere in his account, Narumi comments on high prices for materials, high wages, and the enthusiasm with which members of the construction trades pursued these wages.[73] Many residents rebuilt their roofs with boards or thatch, just in time for the great typhoon of 1856 to blow them away.[74]

Storehouses were associated with the bakufu insofar as the city government dictated their design. They were also associated with wealth for the obvious reason that the wealthiest residents of Edo required large or multiple storehouses. The particularly severe and extensive damage to storehouses, therefore, would have contributed to a sense of the cosmic forces redistributing wealth. There was no looting or other serious civil disorder in the wake of the earthquake, but there was plenty of work at high wages for those who would repair the storehouses and residences of the wealthy.

Patterns of Benefit

The earthquake produced new deities such as Second-to-None Plasterer Buddha (Sakanmuni nyorai) and Roof Tile Earthen Storehouse Bodhisattva (Yane-no-kawara dozō bosatsu), who appeared in a farcical broadside print featuring the pair in the context of a *kaichō*, the periodic revealing of an otherwise hidden Buddhist divinity (fig. 7). To the right of the print is a temple sign announcing the exhibition of the images from the night of the second day of the tenth month, for a daily fee. The storehouse-headed bodhisattva is in the center of the print, holding a key to open storehouse doors of the wealthy in his right hand and a rice ball in his left. To his right a plasterer's assistant points to the manifest deity using a mixing stick. The revealed Plasterer-Buddha stands on a stool. He holds a trowel in this right hand and a plasterboard in his left. Wooden supports prop up his holy enclosure, which shows evidence of damage to its walls. The buddha, of course, is a parody of Shakyamuni, and the bodhisattva is a parody of Jizō

FIGURE 7 Print of Second-to-None Plasterer Buddha and Storehouse Bodhisattva, revealed in a public display of Buddhist icons (1855). Courtesy of the Tokyo Metropolitan Library.

(the "Earth-Storehouse" Bodhisattva). The origin story (Yuraiki) text above the figures cleverly takes the classical story of Shakyamuni's life and modifies it: the Plasterer Buddha hears a voice after the earthquake that says, "There are many plaster walls!" and finds riches after bathing in the water of the city waterworks. He becomes addicted to prostitutes (*yūjo bosatsu*, "prostitute bodhisattva"), however, and ends up penniless.[75] These two deities have arrived to save Edo from the effects of the earthquake, after receiving suitable payment, of course.

Keeping wages and prices under control was an early and ongoing concern of City Magistrate officials. It was also a concern of many of the prominent writers of earthquake accounts. Regarding the likelihood of the cost of materials and workers' wages rising, Jōtō Sanjin explained that the bakufu decreed that workers receive one or two *bu* in wages and that materials such as mats and rope be sold at only 10 to 20 percent above ordinary prices.[76] *Record of the Times* praises the benevolence of our "enlightened ruler" for, among other things, "strictly decreeing" that there be no greed or excess in the price of grain and building materials and in wages charged for repairs and rebuilding.[77] The bakufu, however, was unable to change obvious wage and price dynamics by fiat. In a more realistic view of the situation, Miyazaki Narumi points out that "because skilled workers have been so busy after the earthquake and have received much money, it is only natural that they would spend their money wildly." He continues explaining that merchants are doing a brisk business in towels, belly wraps, and the other kinds of clothing and accessories construction workers use. Elsewhere Narumi mentions that the high demand for roofers, carpenters, and plasterers and their relatively low numbers had produced imposters: "People have been posing as members of these trades and enjoying the receipt of high wages."[78]

In the aftermath of the earthquake, a clear sense of winners and losers developed, which is apparent in several catfish prints. *Tipsiness after the Great Catfish (Ō-namazu-go no namayoi)* depicts a large group immediately after the earthquake, still somewhat disoriented. The Kashima deity vigorously skewers the giant catfish with a sword, and the huge fish divides the print into upper and lower sections. The dozens of people depicted thus divide into two groups. Those at the top are labeled "smiling," while those at the bottom are "weeping" and "have plenty of free time"—that is, they are unemployed. The smiling group includes a carpenter, a plasterer, a seller of lumber, a blacksmith, a roof-tile merchant, an elite courtesan, an ordinary prostitute, a physician, and sellers of certain types of ready-to-eat foods. In total, the print depicts about thirty specific occupations as profiting from the earthquake. The crying group includes a teahouse proprietor; a seller of eels; a variety of entertainers such as musicians, comedians, and storytellers; a seller of luxury goods; a diamond merchant; and a seller of imported goods—twenty-five specific occupations in all.[79] Other catfish prints taking up the theme of society divided along the lines of economic winners and losers portray similar sets of occupations, although the elite courtesan sometimes ends up in the "idle" category.[80] In looking at the mix of occupations on each side of the giant catfish divide, we find a society consisting of interdependent, specialized occupations serving varying income levels.

Although the earthquake has divided this society, it has also brought people together. Prints like *Tipsiness after the Great Catfish* were typically ambivalent. The people on each side of the giant catfish are dressed similarly and assume similar postures. The winners are not celebrating, and everyone appears dazed (despite the "smiling" label, nobody is actually grinning). The earthquake has united them in a common terrifying experience. Certainly the sellers of luxury goods, for example, will suffer from the diversion of money to such basics as building supplies and construction work. There is no suggestion in the print, however, of censure or that those on the crying side of the earthquake deserve any cosmic punishment. Instead, the earthquake is an example of the instability of the world, a basic tenet of Buddhism. Those townspeople harmed deserve compassion and assistance. Indeed, several catfish prints harshly criticize construction workers who have become arrogant, drinking and whoring away their windfall profits while others still suffer from the effects of the earthquake.[81]

Squeezing the Wealthy to Renew the World

Significantly, the suffering townspeople who deserve compassion do not include wealthy merchants, who especially in the catfish prints became deserving objects of cosmic chastisement. A common motif in the prints is the earthquake distributing money from wealthy merchants to ordinary townspeople. Prints typically depict prosperous merchants amidst the ruins of their collapsed storehouses vomiting or excreting gold coins. Often an anthropomorphized catfish prods the afflicted person to emphasize the redistributive role of the earthquake, forcing the rich to vomit and excrete money.

Distinguished Millionaires (*Mochimaru no chōsha*) is a good example, beginning with the title. *Mochi* means to possess, and *maru* (round) can mean money in the form of coins. A common gesture indicating money, then as now, is to make a circle with the thumb and index finger of the right hand. In this print, *maru* is written not with the usual Chinese character but as a circle, with the additional possible meaning of zero. Moreover, the *chō* of *chōsha* is written backwards to indicate a reversal of fortunes. The print features three wealthy or formerly wealthy men in the foreground excreting gold and silver coins. Hovering above them is a nearly destroyed storehouse. The text begins with the storehouse, pointing out that the men lack the power to restore it. The three then begin a conversation, which reveals that in addition to the obvious earthquake damage, they have been pressured to provide charitable aid (*segyō*). The first man, for example, says, "Indeed, indeed, I am greatly burdened. Even if I were to distribute aid here and there, I doubt that I would acquire hidden virtue [*intoku*]."[82] Both "aid" and "hidden virtue" (providing assistance without someone else knowing) are terms with Buddhist overtones, and the speaker doubts that he will get any karmic return on his investment of charitable aid.

A print featuring a neck tug-of-war between a guild head (*zatō*) representing the rich and a giant catfish has the catfish declaring that the earthquake was divine punishment for excessive greed contrary to the teachings of the *kami* and buddhas.[83] In many prints of this type, the recipients of the flow of wealth from elite purses are clearly skilled laborers. Indeed, one print is entitled *Distinguished Skilled Laborer Millionaires* (*Marumochi shokunin chōsha*). Here too, the *chō* of *chōsha* is reversed left and right to indicate a reversal of fortune. The seven human figures in the print, reminiscent of the seven deities of good fortune, form a circle. At the top, two wealthy men excrete money that six laborers pocket. One laborer exclaims, "We have plenty of work and plenty of liquor," all "thanks to Earthquake-sama."[84] Similarly, in the print *Erecting Peace* (*Taira no tatemai*), a crew of anthropomorphic catfish dressed as construction workers, along with the deity Daikoku, work to erect a massive structure that forms the character *taira* (peace, calm). One of the catfish tills the ground beneath the character, which yields gold coins. The six catfish standing in a circle below it, plus Daikoku, are also reminiscent of the seven deities of good fortune. The verse at the top of the print reads, "Mixing together wealth and poverty, the catfish erect a world of peace," a peace that was the result of a redistribution of wealth.[85]

The print *Prosperity Treasure Ship* (*Hanjō takarabune*) contains only a short song as text: "With plans to make big money, we bravely rise up from the waves" (fig. 8). Rising from the waves suggests profit from the recent destruction, and the visual elements of the print make this point in several ways. The usual "treasure ship" image is that of the seven deities of good fortune happily sailing off to Penglai (J. Hōrai), the island of the immortals off the coast of China. *Prosperity Treasure Ship* replaces the seven deities of good fortune with seven skilled laborers, including a carpenter (replacing Ebisu), a general construction laborer (replacing Daikoku), a high-beam construction worker (replacing Bishōmon-ten), a courtesan (replacing Benzai-ten), a roofer (replacing Hotei), a tile vendor (replacing Fukurokujū), and a plasterer (replacing Jurōjin). A giant catfish as the earthquake serves as the body of the ship, and its sail is the cracked wall of a storehouse propped up with wooden supports. The bow of the ship is fire in the form of a flaming dragon. Other standard iconographic elements in treasure ship prints include a pine tree (symbol of longevity), which here is replaced by bundles of copper cash. The crane flying overhead in the print consists of various dried fish products, and the turtle swimming nearby

FIGURE 8 *Prosperity Treasure Ship (Hanjō takarabune,* 1855). Courtesy of the National Diet Library, Japan (http://dl.ndl.go.jp/info:ndljp/pid/1302079).

consists of dried squid. Vendors and suppliers of such ready-to-eat food products prospered in the aftermath of the earthquake. The sea upon which the boat sails consists of blue roof tiles and roof tile fragments.[86]

The Ansei Edo earthquake created a short-term transfer of wealth from one group of commoners—rich merchants—to less affluent commoners, especially skilled laborers. One mechanism was simply that wealthy civilian survivors and government officials needed to rebuild their storehouses and mansions. Less obvious but probably more significant was social pressure, including bakufu pressure, to contribute to the relief effort. In most prints and popular literature, the townspeople used the term "world renewal" (*yonaoshi*) to characterize this wealth transfer.

If popular perceptions of this wealth transfer were accurate, what levels of wage increase took place during the early weeks of the rebuilding? The City Magistrate recognized that wages and prices would inevitably rise, so its strategy was to allow what it regarded as a reasonable increase and prohibit anything beyond that. Noguchi has summarized these official rates in tabular form.[87] For example, the official increase for full-fledged carpenters was 50 percent, from 3.0 *monme* of silver to 4.5. Such an increase would have been substantial, of course, but less than what one might expect from the vast downward flows of gold and silver depicted in the catfish prints. Indeed, actual wages were much higher than the official rates. By the middle of the tenth month, some carpenters were commanding five times their usual wages, and plasterers and roofers were sometimes working at seven times normal. Wages became so high that many domains began bringing their own construction workers to Edo, even from a great distance. The Hiromae domain in the Tōhoku region, for example, stopped hiring local labor and brought in seventy-six workers from its home territory. This influx of workers from the outside may have been the reason for a slight decline in wage levels during the eleventh month.[88]

Sakuma Chōkei, the young police official, was heavily involved in relief efforts. Looking back on the event many years later, he pointed out that City Magistrate officials were worried about unreasonable wages by skilled laborers. In retrospect, however, inevitable market forces were at work. In the years before the earthquake, many skilled laborers had left Edo for their home provinces because the economy was slow. Therefore, they were in short supply even without the intervention of the earthquake. The earthquake provided an opportunity for great profits, which caused skilled laborers from around Japan to journey to Edo. This influx "naturally lowered wage rates." Moreover, Sakuma reminisced, because the earthquake brought renewed prosperity to Edo, we should look upon the activities of these workers in a positive light.[89]

Although official attention and popular prints focused on the profits of skilled laborers, we should bear in mind that unskilled laborers and many other occupations were in a position to benefit economically from the earthquake. Nearly any able-bodied man, for example, would have been able to find work as a hauler of dirt or debris during the rebuilding. Indeed, at one point wages for dirt haulers reached ten times their usual rate. Similarly, porters of all kinds were in demand. Food was abundant, thanks in large part to the recent completion of a bountiful harvest that year. Sellers of rice cakes, dried fish, and other ready-to-eat foods, and vendors of small items such as towels or footwear would have been among the many common people for whom the earthquake ended up providing temporary financial relief.[90] The earthquake started as a terrifying disaster, but for many of Edo's townspeople it soon became a source of prosperity.

Earthquake as Deity

For a few months, human entities represented by such figures as Second-to-None Plasterer Buddha and Roof Tile Earthen Storehouse Bodhisattva enjoyed elevated status and income. "Earthquake-sama," of course, made this all possible. It was therefore logical, at least for those who had benefited from it, to deify the earthquake. Earthquake-sama became a deity-of-the-moment (*hayarigami*), represented as a giant catfish. The print *Catfish Hanging Scroll* (*Namazu no kakejiku*) shows an image of a giant catfish painted on a hanging scroll, attended by both a shrine and a temple priest. The throng of people venerating the catfish consists of occupations that benefited from the earthquake, including a plasterer, used clothing merchant, bar (*izaka*) proprietor, roofer, bone healer, lumber merchant, carpenter, hardware merchant, regular construction worker, rickshaw puller, and a brothel proprietor. The Foundation Stone that normally pins down the great beast has become a small, snail-like adornment on the head of the catfish.[91]

In *Joyous Self-Precaution* (*Manzairaku mi no yōjin*), the earthquake catfish similarly appears as an image on a hanging scroll. Various professions such as carpenters and other construction workers, a hardware vendor, and a footwear vendor are asking the earthquake catfish to look after them in the future. The passage ends with a seller of outbuildings saying, "Praise,

praise, praise . . . [*namu*]!" followed by a variation of the expression we have seen repeatedly: though the earth may shake, the Foundation Stone remains in place. The text of the print, however, leaves out any mention of the everlasting reign of the sovereign.[92] This merchant is not concerned with such abstractions. He and those around him venerate the earthquake catfish, owing to the concrete wealth it has brought them. In prints such as these, we see that the earthquake itself took on a divine quality in the eyes of some of the townspeople, the logical culmination of interpreting it as an act of cosmic world renewal.

Patterns of Relief

"The state," began a passage in Saitō Gesshin's account of the earthquake, "has decreed that for the sake of obtaining repose for all the wandering spirits resulting from those who have perished in the present calamity, rites to succor starving ghosts will be conducted at the following temples on the second day of the eleventh month. Each of these temples will receive fifteen pieces of silver and a grant of an additional ten pieces."[93] The next entry in Gesshin's account lists twelve temples and a few other details.[94] By requiring this religious event at the one-month anniversary of the earthquake, bakufu officials likely intended to create what today we would call a sense of closure. The decree itself enjoins everyone in the city "without exception" to observe the solemnity of this event.[95] During the previous month, entities within the bakufu had been busily engaged in dealing with the results of the calamity.

Bakufu Assistance to Its Retainers

The bakufu was primarily a military organization, and the bulk of its resources went to providing relief for stricken retainers. On the night of the earthquake, the overwhelming response of daimyō and other bakufu retainers was to make their way to Edo Castle to inquire about the well-being of Shogun Iesada, who had escaped to the Fukiage Garden within the castle along with his family.[96] The day after the main shock, the bakufu sent small teams of messengers to each daimyō mansion. The format was that of an inquiry about the daimyō's well-being, but these visits also served the purpose of a preliminary gathering of information about daimyō conditions. The senior councilor on duty on the third day was Kuze Hirochika, who had temporarily replaced Abe Masahiro because Abe's residence had collapsed. Somewhat oddly, Hirochika sent an order that all daimyō in the city report that evening to inquire on the shogun or to send a representative should they be physically unable.[97] This order clearly irritated many of those who had risked their lives and neglected their own households to visit the castle the previous night. This irritation forced bakufu officials verbally to excuse most daimyō from attending on the third, and only twenty-four showed up in person (the others sent messages). Realizing the desperate condition of many *fudai* daimyō households, the bakufu issued a directive on the fourth day permitting daimyō to return to their domains at their convenience.[98] On the seventh day, however, perhaps realizing that coastal defenses would be unmanned, the bakufu partially reversed itself. It decreed that all daimyō whose residences were salvageable should strive to remain in Edo.[99]

This flip-flopping was a surface manifestation of conflicting pressures behind the scenes. To take one example, Abe Masahiro advocated a generous policy toward daimyō, and he stayed away from Edo Castle on the seventh and eighth days to attend to his devastated household. He returned on the ninth day only after Hotta Masayoshi was appointed chair of the senior councilors, thus taking on many of Abe's duties. Mito Nariaki, whose household also suffered many deaths and severe destruction, was opposed to any leniency in standards. On the tenth day, Nariaki sent Masahiro a demanding letter, chastising him for allowing the daimyō to return to their home territories on account of the earthquake. The letter suggested that such a move would start a trend whereby daimyō would seek to be excused from their attendance duties.[100] In light of the relaxation of attendance requirements in 1862, Nariaki may have been prescient on this point.

The major source of bakufu aid to daimyō and other retainers came in the form of loans on generous terms, which the bakufu could ill afford to provide. The loans were at zero interest, repaid in installments over ten years. Authorization of loans began as early as the fourth day, and the order in which retainers received loans was based on the importance of their duties. Loan amounts ranged from 10,000 *ryō* in the case of some senior councilors to 2,000 *ryō* for lesser daimyō officials. The total amount disbursed to daimyō was 61,000 *ryō*. It was on the seventh day that bakufu officials decided on measures for lesser retainers, the bannermen and *gokenin*. The bakufu made a distinction between those with territories of between one hundred and ten thousand *koku*, who received ten-year, zero-interest loans. Those receiving direct stipends of one hundred *hyō* (bales of rice) or less received outright grants. Grants or loans were based on property damage only, not on the human toll. The total amount dispersed to bannermen of five hundred *koku* or higher was 89,177 *ryō*, with a total of 1,658 retainers receiving funds. [101]

Bakufu Assistance to the Townspeople

While acknowledging that the relationship between the bakufu and ordinary residents of Edo was different from that of a modern state, Noguchi Takehiko nevertheless criticizes the bakufu for disproportionately aiding its military retainers both in terms of sheer quantities of money and administrative resources. As the "public" (*ōyake*) authority, he claims, the bakufu should at least have given the appearance of supporting the general prosperity of the realm. The Machigaisho, the major entity for providing civilian relief, was well financed in 1855 owing to a bountiful rice harvest. This good harvest also helped keep food prices—and therefore warrior incomes—low. Nevertheless, the Machigaisho was still inadequate for the task of relief of the general population and had to rely heavily on private charity.[102] Noguchi seems to underestimate the point about the bakufu not being a modern state. Today we unrealistically expect states to be almost godlike in their ability to mitigate natural hazards. The townspeople of Edo, however, did not expect the same degree of government aid that the people of, for example, Kobe did in 1995. Spared any serious physical destruction, the City Magistrate and Machigaisho began their work even faster than other bakufu entities—the very night of the earthquake—and leveraged private charity effectively to multiply their own resources. On balance, relief to the townspeople was a case of successful disaster management, especially by the standards of the time.

City Magistrate officials planned initial relief efforts the night of the earthquake, and the next morning they issued the following set of directives to the neighborhood heads:

1. Distribute rice balls to the disaster victims.
2. Set up temporary shelters where the homeless have congregated.
3. Render speedy aid to the wounded.
4. Summon the heads of the wholesale distributors and have them secure and stockpile daily necessities and items in great demand.
5. Order the heads of trade associations to bring skilled workers from the countryside into Edo.

6. Prohibit sellers from holding items back from the market and buyers from cornering the market.
7. Control price and wage increases.
8. Order police officials to patrol the city, rendering aid and enforcing regulations.
9. Assign emergency assistance duties to the neighborhood heads.[103]

In practice, it was not possible to pursue methodical relief activities until after the fires had been brought under control on the fourth day. Systematic rounds by City Magistrate officials began on the fifth day.[104]

The first order of business was dealing with the dead and injured, a topic already discussed with respect to surveys, casualty statistics, and popular perceptions. The City Magistrate ordered Ekōin in Honjo to serve as the place to bring unclaimed bodies. It also ordered a supplemental survey of serious injuries on the ninth day (turned in on the eleventh).[105] Regarding the wounded, the Machigaisho established treatment stations and recorded the names of as many of the victims as possible. For example, a famous bone doctor named Nakura lived in the Ryōgoku area, and the road and area in front of his house became a sprawling triage station.[106] The case of a bakufu-supported hospital for the poor provides some longer-term perspective. The Koishikawa Yōjōsho, a charity hospital, was established in 1722 and reorganized in 1843. A notice sent to the neighborhood heads issued on the tenth day of the twelfth month states that applicants to the hospital have lately become few in number. The notice goes on to say that neighborhood heads should locate impoverished victims of the recent earthquake who may be receiving medical treatment in other circumstances and urge them to make use of the benevolent services of the hospital.[107] This notice suggests that roughly two months after the main shock, there was more than enough capacity to treat earthquake victims who continued to require care.

The second emergency relief policy was setting up temporary shelters and distributing cooked rice. Even people who might ultimately have benefited from the earthquake initially faced many hardships. For example, a ban on fires and a lack of cooking utensils meant that even if rice or other grain was at hand, there was no way to cook it.[108] The Machigaisho dealt with major fires in 1806, 1829, 1834, and 1845, as well as the Tenpō famine of the 1830s. When the earthquake struck, it was able to erect one thousand *tsubo* (about 0.33 hectares) of temporary dwellings in half a day owing to prior experience in providing fire relief and to its connections with civilian contractors. The Machigaisho sent out public notices about temporary housing on the fourth day, initially in three locations.[109] It ultimately established housing in five locations, three of which were operating by the sixth day, with two more set up by the thirteenth day. The modular dwellings were made mainly out of wood, with most of the needed materials available in warehouses. Sakuma Chōkei likened this work to putting up tents.[110]

Seeking to make the most of its modest resources, the Machigaisho was careful not to duplicate relief assistance. It first distributed ready-to-eat food to the needy inside their houses. Later, each household sent a representative to a collection point to pick up its share of uncooked rice.[111] Those who entered temporary housing received food aid there and were not eligible for other food handouts. As officials made their rounds to determine which households needed food assistance, they also determined which houses were so badly damaged that their residents were eligible for temporary housing. According to Sakuma, several City Magistrate staff filled a stretcherlike device with food and set out to distribute it. Assisted by neighborhood heads, the retinue hoisted an official duty flag for easy identification. Landlords assisted in transporting relief supplies and distributing them to their tenants. The city came to life in a flurry of activity as people rushed around to look after neighbors or acquaintances, provide aid, or transport the injured to medical stations.[112]

About 202,400 people across Edo received cooked rice, the distribution of which stopped on the twentieth day, replaced by handouts of uncooked rice.[113] The conditions established for those continuing to receive food aid were based on income level or poverty, not earthquake damage. Examples of eligible categories included "workers supporting a family on day wages," "subcontractors to skilled workers," "shopkeepers whose sales are insufficient," or "landlords with holdings so small that they have to do outside work to make ends meet."[114] The population of day laborers in Edo in 1855 numbered about 288,000, and ultimately about 381,200 people received relief rice of some kind for days or weeks after the earthquake. In other words, the bakufu conducted large-scale poverty relief under the banner of earthquake relief. Moreover, this situation occurred earlier, following the same pattern as recent large fires or the temporary shelter set up during the Tenpō famine of the early 1830s. During the nineteenth century, townspeople had become an explicit object of government relief efforts.[115]

It is possible that this policy of aiding those of modest means resulted at least in part from fears of social unrest. Riots prompted by food shortages had rocked Edo and all of Japan's other major cities in 1787, one reason for the establishment of the Machigaisho shortly thereafter. When riots broke out in many other parts of Japan in 1837, Edo remained calm in part because of the ability of the Machigaisho to provide relief.[116] Massive food aid was surely a major reason that there was no looting, rioting, or mob violence in 1855, in contrast with 1923.

Control of wages and prices was an area in which the bakufu was generally ineffective. At least in the minds of bakufu officials, control of prices was closely connected with the suppression of rumors, another notoriously difficult task. For example, on the fifth day the City Magistrate ordered city officials to suppress and correct rumors that the bakufu had ordered a business holiday when in fact the official policy was to encourage business to operate. Moreover, officials were to suppress rumors of food shortages.[117] In the realm of wage increases, the City Magistrate posted notices, had notices read to workers directly, and forced work crew chiefs to affix their seals acknowledging rules concerning wages. Its officials even arrested a few offenders here and there.[118] Decrees regarding wages and prices outnumber those about any other topic. [119] The City Magistrate, however, could not change market forces by decree, nor could it afford to arrest significant numbers of those who were essential for rebuilding the city. One decree from the twenty-third day of the twelfth month, almost two months after the main shock, can serve as a benchmark. It complained of "illegal" prices for lumber and construction wages and threatened punishment (without specifics) for anyone investigators might find guilty of violating previous decrees.[120]

When not dealing with emergencies, the Machigaisho made loans in the manner of the "righteous granaries" (gisō), de facto banks that had become established in many domains during the Tokugawa period.[121] Owing to the earthquake, beginning at the end of the twelfth month and extending into 1856 the Machigaisho issued edicts suspending repayments on outstanding loans for periods ranging from two to twelve months, depending on the degree and nature of damage borrowers had sustained.[122] In a related realm, the City Magistrate tried to deal with false rumors concerning money. For example, on the fifth day of the eleventh month, a directive explained that earthquake-related rumors are "not few" and that some people had become so confused that they stopped using money, thus hindering its circulation. The directive strictly prohibited dissemination of rumors.[123] Although it did not specifically mention unauthorized prints, the same directive became the basis for an eventual crackdown on the production of catfish prints. Some of the alleged rumors have a contemporary ring to them. For example, on the twentieth day of the tenth month the City Magistrate issued an edict denying a rumor that all debts had been canceled by government order and warned against con artists claiming to be able to secure bakufu loans.[124] In many respects, public behavior during the weeks after the earthquake was exemplary, but rumors and deceit flourished in the post-earthquake confusion.[125]

Daimyō and Religious Institutions Assist the Townspeople

Governing the city of Edo was the prerogative and responsibility of the bakufu. Therefore, warrior houses generally provided aid discreetly and only after the City Magistrate began to solicit charitable donations from the public. Prior to the Ansei Edo earthquake, the leading warrior houses had not been active in disaster assistance, although some of them did aid victims of the Tenpō famine and large fires indirectly. In such cases, unemployed samurai or female members of the warrior household gave donations as private individuals, but the warrior household's treasury was the real source of the charity.[126] Following the earthquake, many warrior houses took the unprecedented step of issuing charitable aid directly.

One immediate form of relief was for daimyō whose mansion grounds were in reasonably good condition to allow homeless commoners to camp there temporarily. As the scale of the disaster became known, warrior houses not pressed by their own problems began to provide aid directly to needy areas. According to *Ansei Chronicle,* the Sendai and Tomiyama domains gave the largest quantities of aid. Praising the lord of Sendai for his virtue, *Ansei Chronicle* explains that he authorized the distribution of money and rice to devastated townspeople in four nearby neighborhoods.[127] The popular press frequently reported on the contributions of warrior houses but typically omitted their names owing to the sensitive nature of such donations. Sendai, however, made no attempt to remain anonymous in its giving. It sent ten thousand bales of rice to the bakufu for use in disaster relief and provided extensive food aid to the houses and businesses in the areas surrounding its mansions. In all, the domain provided approximately one bale of rice for each of over 425 structures in the zone. The Tomiyama domain distributed over 130 *koku* of rice. Other domains made lesser contributions, and several leading bakufu officials made significant contributions from their own resources.[128] One characteristic of this domain-derived aid is that it tended to go to districts and neighborhoods receiving relatively little from the donations of wealthy townspeople, suggesting effective coordination of aid distribution.[129]

Buddhist temples were another source of disaster relief, both tangible and intangible. On the second day of the eleventh month, as we have seen, the bakufu paid thirteen large temples to perform rites for assuaging starving ghosts. The material assistance temples provided consisted mainly of temporary shelter for the homeless and aid to their possessions outside the temple grounds. Some also engaged in large-scale charity.[130] The Shibazōjō Temple, for example, distributed 250 *ryō* in relief money. As a reward for this generosity, the shogun personally gave the head priest a set of seasonal attire.[131] The bakufu also recognized several other temples in formal announcements and awards.[132]

Townspeople Helping Townspeople

Some townspeople of relatively moderate means were generous in providing relief. The bakufu actively rewarded those who provided aid, typically by posting their names and amounts at each neighborhood's public notice area and often bestowing monetary tokens of recognition. To cite but one example from official documents, a declaration dated the twenty-seventh day of the tenth month states that Fukagawa householder Mataemon distributed relief in the form of small amounts of cash to a total of 159 people in fourteen neighborhoods other than his own, for a total contribution of 34 *ryō*. For his deeds, the bakufu formally recognized Mataemon with two pieces of "reward silver."[133] For many engaged in business enterprises, such aid was a form of social obligation. At the same time, it was a way of displaying one's de facto social standing.[134] Often when the discussion in *Ansei Chronicle* turns to a particular district or neighborhood, it presents a list of names, amounts, and occasionally other details of charitable contributors from that place.[135]

Following a disaster on a scale of the earthquake, both commoner society and the bakufu expected wealthy merchant houses and large organizations to provide relief from their vast resources. The possibility of mob violence such as that which occurred in 1787 undoubtedly enhanced this social pressure. An examination of donations by wealthy commoners indicates that in many instances, their giving was coordinated and planned. In one case, for example, five wealthy merchant houses from the same neighborhood each donated the same amount. They surely decided on a total figure and then divided the burden equally. In other cases, donations were the result of discussions between wealthy commoners and local government officials. Expectations of good social behavior served as a quasi-coercive mechanism for generating an outpouring of private charity.[136]

Kitahara examines several cases in which self-interest was clearly the motivation for providing earthquake relief. Mitsui, for example, operated a clothing shop and a money-changing shop in Surugachō of Nihonbashi. These businesses made many donations, but they were mainly to suppliers, subcontractors, and other smaller businesses that performed work essential for Mitsui's welfare. Mitsui provided a sizable sum of 277 *ryō* for disaster relief. It received no formal praise from the bakufu, however, probably because only 56 *ryō* went to those with no business relationship with Mitsui.[137] Less obviously self-serving were cases in which businesses that stood to profit handsomely from the earthquake made generous charitable donations to the needy. A hardware guild donated 200 *ryō*, for example, and the Fukagawa Lumber Association donated 750 *kanmon* of copper cash.[138] Such donations, of course, made good sense from a public relations standpoint, while also providing genuinely needed assistance.

Despite a self-serving quality to much of the charitable donations by wealthy merchants and business organizations, statistical and anecdotal evidence indicates a large number and a wide variety of private contributions, the majority of which were modest donations by ordinary townspeople. There were 174 instances of donations to the temporary shelters of such things as rice, pickled plums, pickles, and pots, and 255 people made cash donations to the city's poor that totaled 15,037 *ryō*. Similar to wealthier donors, there was a strong tendency for people to give aid to their neighbors or to their tenants. Relatively wealthy neighborhoods, therefore, tended to be richer in charitable aid. Using the same districts as for population statistics, the range of charitable donations generated from a district's own residents ranged from 13 to 3,733 *ryō*. This large disparity suggests one reason the bakufu distributed aid based on relative poverty and income levels.[139]

Those who could not contribute money often provided goods or services. A partial list includes bean paste, tea, noodles, pickled radishes, pickled plums, sweet potatoes, dried fish, straw mats, towels, paper, pickled vegetables, and other practical items. One hair dresser offered free services. Donations were not limited to the residents of Edo. A rural physician donated two hundred packages of medicine for treating cuts, bruises, and puncture wounds, and the peasants of one rural village donated six barrels of pickles. Donations of useful goods and services rarely registered in official statistics, but such contributions were significant. All indications point to a strong spirit of mutual assistance, and the Ansei Edo earthquake was not the first manifestation of widespread charity in the city. There was a similar reaction to previous disasters from at least the time of the Sakumachō fire of 1829.[140]

Charitable giving was part of normal life in nineteenth-century Edo. With bakufu sanction in 1822, the heads of *hinin* (officially recognized beggar) groups began periodically collecting a small sum from each household in return for promises of *hinin* cooperation in suppressing the activities of unsanctioned beggars. In short, "giving" in ordinary times had become a de facto tax, one that reinforced differences in social status. Charitable giving after the earthquake, however, took on nearly the opposite quality, which Kitahara characterizes as *"girei-teki."* Although this word normally brings to mind rigid formalities, what she means in this context is something akin to a theatrical performance whereby that which is visionary or otherwise impractical is temporarily enacted within altered social reality. One reason for the widespread outpouring of post-earthquake charity is that it complemented the earthquake itself in destroying or blurring ordinary social distinctions. Such giving thus had a psychologically liberating effect, more so for givers than for receivers. In this sense, the earthquake and the charitable giving by ordinary people possessed many of the same qualities and appeal as the spontaneous mass pilgrimages to the Ise Shrine (*okage-mairi*).[141] This insight into the temporary liberating effect of the earthquake may also help explain why so many commoners tended to interpret the event as an instance of social renewal.

Patterns of Rebuilding

Rebuilding and recovery began soon after the main shock of the earthquake. The many daimyō with destroyed or heavily damaged mansions were eager to repair them as quickly as possible, in part for pragmatic reasons and for the sake of appearances. Plasterers, carpenters, and others were eager to partake of this rebuilding and reap the high-wage rewards. One small illustration of this phenomenon comes from Miyazaki Narumi, who tells of a carpenter who left his own destroyed house, wife, and child to make money working on the mansion of the Sakai daimyō. The carpenter received 100 *ryō* and a rice stipend, but Narumi is ambivalent in assessing the carpenter's actions, regarding them as evidence of "lighthearted wisdom."[142] As we have seen, Sakuma Chōkei explained with the benefit of hindsight that this high-profit environment gradually returned to normal owing to market forces. What about the long-term rebuilding of Edo? When, for example, was the city back to the way it was before the earthquake? Most studies of the earthquake have neglected this matter, in part because it can be assessed only indirectly. In two different studies, Kitahara has attempted to answer these questions using documents indicating the status of rebuilding projects from the City Magistrate and from the Treasury Office (Kanjōbugyō).

The earthquake dislocated the wooden troughs that carried water through the Tamagawa Waterworks and the bamboo pipes that connected them. The worst leakage was at Yotsuya-ōdōri because of a gap between the pipes and the stone trough. Several popular prints depicted the water leakage in this area. On the tenth day of the tenth month, the agency operating the waterworks (Fushinbugyō) dispatched a carpenter crew chief to Hanemura. It is not clear what he or his crew did, but the city paid travel expenses until the end of the eleventh month. In the second month of 1856, the water was stopped for five days, and soon thereafter repair work on Edo's wooden troughs began. Repairs continued through at least the twelfth month of 1859. Because the person in charge of earthquake-related construction, including the waterworks, received an award from the bakufu in the twelfth month of 1856, it is likely that all or most essential construction and repair was completed by then—even if complete reconstruction dragged on for several more years.[143] Reconstruction of administrative buildings and the duty residences of major officials began in the tenth month of 1855 and continued until the twelfth month of 1859. Gates and bridges were repaired at a rate of at least one per month, and the whole process took until the tenth month of 1859. Essential repairs were competed by the fourth month of 1857.[144]

One complicating factor in assessing construction progress was the severe typhoon that tore through Edo ten months after the earthquake. It undoubtedly slowed earthquake reconstruction or caused some work to be redone. Summarizing the larger reconstruction picture, the first three months after the earthquake was a time of emergency measures. Early in 1856, systematic rebuilding began in earnest, and the essential work was mostly finished by the end of that year. During 1857, the pace of rebuilding slowed, and some projects dragged on until the end of 1859.[145]

Conclusions

The Ansei Edo earthquake shook the shogun's capital, which was among the largest cities in the world. The earthquake did not focus its destructive fury on downtrodden commoners, as Markus and Clancey have suggested, nor did it take aim mainly at the bakufu, as Noguchi has suggested. Patterns of destruction were mainly a function of the nature of the soil base, and areas built on unconsolidated fill included commoner neighborhoods, prime bakufu and warrior real estate, the offshore artillery batteries, and a prominent playground of the rich, Shin-Yoshiwara. The earthquake occurred at a time when the bakufu, while still the supreme military authority in Japan, was beginning to weaken in ways that would ultimately lead to its demise. The disaster imposed serious financial strain on the bakufu, especially insofar as it was obligated to provide generous relief to its retainers. Bakufu grants and zero-interest loans mitigated the financial strain on retainers and *fudai* daimyō. Most *tozama* daimyō (the "outside lords") resided in areas of minimal damage. The bakufu also took the unprecedented step of temporarily relaxing daimyō attendance requirements. Therefore, the earthquake had the overall effect of weakening the bakufu vis-à-vis the daimyō. The willingness of some daimyō to launch high-profile relief efforts may also have reflected relative bakufu weakness.

The bakufu, or more specifically the City Magistrate and Machigaisho, performed well in directing the relief effort for townspeople. The earthquake did not cause any serious outbreaks of violence, nor did it become an occasion for popular anti-bakufu sentiment. Dealing with the earthquake was probably the last major challenge that the bakufu handled expeditiously. Nevertheless, a closer examination of the psychological impact of the earthquake reveals that it played a role in undermining bakufu power beyond the obvious drain of financial resources. The next chapter examines several less obvious effects of the earthquake on the political balance of power.

Popular discourse quickly labeled the earthquake an instance of *yonaoshi*—world renewal or world rectification. Indeed, the Ansei Edo earthquake was the first seismic upheaval characterized in this way. Despite the grandiose implication of the term taken literally, in the majority of instances "world renewal" referred to a short-term transfer of wealth, mainly within the ranks of the townspeople. One important factor was the devastation of *fudai* daimyō mansions, which pushed up the demand for skilled labor. In this milieu, members of the construction trades, those who sold items or services that supported the construction trades and rebuilding effort, and even unskilled laborers enjoyed windfall profits. These profits were the essence of the typical characterization of the earthquake as world renewal. However, the earthquake did produce another more ominous sense of world renewal to be examined in the next chapter.

CHAPTER 5

Meanings

A catfish print entitled *How the Earthquake Modified Ōtsue Songs* (*Jishin dōka Ōtsue bushi*) explicitly brings the genre of Ōtsue, paintings made for travelers in the vicinity of Ōtsu on the shores of Lake Biwa, into the context of the Ansei Edo earthquake. The printmakers of 1855 frequently appropriated Ōtsue motifs such as a catfish pinned down by a gourd. The top of *How the Earthquake Modified Ōtsue Songs* features Ōtsue verses modified to reflect post-earthquake conditions, and the bottom features images of the major Ōtsue motifs, similarly modified to reflect post-earthquake Edo. The prominent gourd-catfish in the middle says, "From Shinano I worked my way up to the capital along the Tōkaidō," referring to the Zenkōji and the Ansei Tōkai/Nankai earthquakes and possibly others. One of the members of the construction trades holding down the gourd says, "Getting this earthquake under control will renew society."[1] More than any of its predecessors, the Ansei Edo earthquake was a transformational event. The obvious items in this example are the appropriation of Ōtsue, not previously associated with earthquakes, and the transfer of wealth associated with world renewal. The Ansei Edo earthquake occurred during an *okage* year of special religious significance after a series of other major earthquakes that appeared, geographically, to have started in Kyoto (1830, another *okage* year) and worked their way toward the shogun's capital. In such a perspective, denizens of Edo and other Japanese could reasonably regard Ansei Edo as a capstone event, although at least some feared that it was not. Recent earthquakes and other significant events of the recent past took on renewed meaning with the benefit of hindsight at the end of 1855. The Ansei Edo earthquake was rich in meanings, both soon after it occurred and for decades thereafter.

Major earthquakes can function simultaneously to cause social change, as a catalyst accelerating trends already in play, and as a window through which to view society at a time of unusual stress and upheaval. Focusing on popular discourse, this chapter examines ways that the townspeople of Edo found or created meaning in the earthquake. To bring out the broader significance of the earthquake, where appropriate I discuss recent past events such as the arrival of Commodore Matthew Perry and events that occurred twelve years later, especially the outbursts of frenzied dancing known as *ee ja nai ka.* As one short verse circulating at the time put it, "The great earthquake that has shaken and corrected this imbalanced society / Has shaken those above and has shaken those below."[2] My main argument is that although the Ansei Edo earthquake was not a revolutionary event in and of itself, some of its effects conditioned Edo society for major change.

Strong Medicine

The idea of the earthquake as social correction lent itself to medicinal metaphors. The earthquake became strong medicine to cure an ailing society by bringing it back into balance. *Catfish Powdered Medicine Pouches* (*Furidashi namazu-gusuri*) is a good example. The print features an anthropomorphic catfish as a peddler of powdered medicine "made in Giant Gourd Hall." He carries a box with a gourd emblem on it, a fan with "powdered medicine" (*furidashi*) written on it, and a straw-tipped pole. Instead of medicine pouches, miniature figures representing occupations that benefited from the earthquake populate his pole. The text of the print explains that this wondrously efficacious catfish medicine, not to be confused with eel medicine, is effective in restoring health by curing seasonal lack of circulation to enhance the economic vitality of society. It does its work by causing strong movements that dislodge the accumulations in storehouses that have built up over the years. The result is a downward flow of gold and silver that is now well circulating in society. This circulation warms cold-hearted people and cures the disease of poverty. The medicine also cures laziness and excessive luxury. The main dose was taken on the night of the second day of the tenth month, and smaller doses (aftershocks) have been taken thereafter.[3] The earthquake, in other words, functioned as a violent purgative cure for the constipated body of society.

Another print entitled *Wondrous Efficacy of the Earthquake's Blessing* (*Jishin myōsaku kudokusan*) takes a similar approach. It features the

Kashima deity as a medicine seller on a stage pointing to a picture of a giant catfish. The text likens the fires that broke out in thirty-six places after the earthquake to moxa treatments. There was pain at first, caused by dwelling outside in makeshift shelters. But the pain of this treatment yielded ample rewards in the form of abundant work, high wages, and plenty of temporary brothels. The temporary housing in five locations was also a beneficial after-effect. The text then lists benefits deriving from the earthquake, including the use of money and its circulation, debt forgiveness, wealth disgorged and flowing downward, and increased wages—all economic matters. Bad effects of the medicine include damage to houses, storehouses, and the earth's surface and the calling in of loans. Those using this medicine should take it after 10 p.m.[4] In this view, the earthquake was a powerful corrective for society, effective but with severe, painful side effects.

Characterizing the earthquake as medicine to correct or renew an imbalanced society was close in meaning to regarding the event as a divine punishment. As with past earthquakes, those otherwise critical of social morality would have claimed any major earthquake as an affirmation of their agendas. Writings to this effect were marginal, however, and outside of that genre, the theme of divine punishment was relatively muted in 1855. Indeed, this theme was more prominent in the discourse following the 1923 Great Kantō Earthquake.[5] Nevertheless, the idea was surely on the minds of many townspeople, and it occasionally received explicit discussion. Early modern moralists tended to stress that major social groups—but not the ruler (emperor, shogun, or daimyō)—had become greedy, selfish, lazy, and so forth. Because the Ansei Edo earthquake took place during an *okage* year, soon after the signing of the Treaty of Kanagawa, shook the shogun's capital, and destroyed the artillery batteries and other manifestations of bakufu authority, it was almost ideally situated to be read as an anti-bakufu event. On the other hand, the bakufu relief response was fast and effective, and many townspeople enjoyed windfall profits. Moreover, Edo was in effect a company town, with the bakufu as the source of people's livelihood. These factors, much more than any coercive force the bakufu potentially wielded, worked against reading the earthquake as a strong rebuke of the shogun's government. Nevertheless, there were serious political issues at the time, especially in the realm of foreign policy, and many townspeople paid close attention to them. In this context, it is inevitable that some would interpret the earthquake, at least in part, as a rebuke of certain policies. Advocates of the policy of "expelling the barbarians" (*jōi*), for example, did not fare well.

Fujiokaya Diary (*Fujiokaya nikki*) includes earthquake commentary in the form of popular drama, chanted tales (*chobokure*), and political graffiti (*rakugaki*). One fictionalized reworking of a chanted tale, "Rites for Preserving the Thousand-Generation Reign" (Sendai hokyū rei), suggests that the earthquake might be a cause for concern with respect to regime continuation. The passage begins with a lament directed at "Mito-sama" for his role in causing the present earthquake. The deaths of Fujita Tōkō and Toda Hōken, prominent advocates of the expulsion of foreigners and associates of Mito Nariaki, were widely reported in the days after the earthquake.[6] Moreover, Mito Nariaki had recently incurred the enmity of Buddhist temples owing to his formal proposal to the bakufu, ultimately rejected, that temples contribute their bells to be melted down to cast cannon for coastal defense.[7] The passage next explains that "indeed, it must be divine punishment for the earthquake to have occurred while the august deities were all away to Izumo" and implores Mito-sama to think about this matter well. The passage goes on to suggest that the earthquake was ultimately the "divine medicine" of Ise. Although primarily a reference to Amaterasu, here "Ise" probably had a secondary association with bakufu leader Abe Masahiro, whose court title was "Ise no kami" (Lord of Ise). Although Abe and the other bakufu senior councilors ultimately rejected Nariaki's proposal, they did consider it or at least gave the appearance of considering it. The text then asks "Ise-sama" why, despite his wise compassion, he allowed Edo to be shaken up to an extent not even likely to occur in China? It abruptly changes tone in the end, praising the establishment of temporary shelters as "benevolent compassion and succor that has soaked people with tears of gratitude." The passage concludes by characterizing this situation as an instance of world renewal.[8]

Nariaki's proposed melting down of temple bells was widely known in the months prior to the earthquake, and it produced political graffiti. One example reads, "Arrogance not corrected as the years go by—crushing temples to forge cannon." Significantly, there was an upsurge of this kind of graffiti in the wake of the earthquake. The typical approach was to blame Nariaki for causing the earthquake by incurring divine wrath. This example refers to Kashima's absence during the tenth month: "As punishment, Heaven shook the earth, caused the deity to be absent, crashed houses to the ground, and froze the marketplace."[9] A catfish print also portrayed the earthquake as divine punishment for Nariaki's proposal. Nariaki sits menacingly atop a temple bell, rifle in hand, while the spirits of the dead cluster behind him. A figure representing Abe Masahiro hides inside the bell. In this portrayal, the earthquake and all of its consequences are the responsibility of Nariaki and possibly the cowardly Masahiro for not firmly rejecting Nariaki's proposal to begin with.[10] Clearly, Nariaki's opponents used the occasion of the earthquake to advocate for their views. Moreover, it is possible that such advocacy, combined with the deaths of expulsion advocates Tōkō and Hōken, helped sway the views of some townspeople regarding pressing issues of the day.

These items indicate some of the complexities in interpreting the Ansei Edo earthquake in political terms. The situation on the ground was multifaceted, in terms of patterns of destruction, different interests within society, and different political factions. There was a rough consensus among commentators at the time that the earthquake was a blow to expulsion advocates in the realm of foreign affairs. Moreover, there was a tendency at many levels of society to criticize Nariaki, whose household and mansion suffered severe damage and a high death toll, the stress from which seems to have adversely affected his psychological stability. Although Nariaki was a member of the Tokugawa family, he and his policies were out of favor when the earthquake struck. It is possible to find implied critiques of the bakufu, with deities serving as proxies. One example is a broadside featuring Amaterasu arriving in Edo (fig. 9). Called "Tentō," Amaterasu appears in human form, clad in priestly robes, whose head is replaced by the shining sun. In modern Japanese, "Tentō" refers to the sun in an anthropomorphic way (usually as O-Tentō-sama), and in premodern texts the name often referred to a supreme ruler of the universe. In this print, Tentō is clearly the equivalent of Amaterasu. A crowd has gathered, some of whom worship Tentō, while others attack Kashima with their fists and push over the Foundation Stone. Two Kashima shrine priests try in vain to restrain the angry mob. Tentō carries a hoe, and gold coins appear in the ground he digs up with it. In the text, the crowd states that there is no need for the Foundation Stone.[11]

Kashima was the highest protective deity of Edo, who guarded the dangerous northeast approach to the city. The bakufu supported the Kashima Shrine as the outermost ring of a network of supernatural forces defending the shogun's castle.[12] The potential anti-bakufu message is clear from the visual components in the print: to prosper, replace Kashima (bakufu) with Amaterasu (imperial court), violently if necessary. The message in the print is especially powerful in light of the frequent references elsewhere to the Foundation Stone as guarantor of social and political continuity.

The revolutionary potential expressed in this print did not occur in 1855, nor did it even become a prominent theme in rhetoric. Indeed, the bakufu itself, especially the City Magistrate and Machigaisho, came through the earthquake looking very good in the eyes of many townspeople. Temporary housing in particular stood out as an example of benevolent government. For example, *Record of the Times* (*Jifūroku*) praises the "august benevolence" of the "wise lord" (*meikun*) for saving the poor through food relief and temporary shelters.[13] *Fujiokaya Diary* contains a short set of Buddhist-style associations. High prices, for example, are associated with "greed," and the temporary shelters are associated with "benevolent government."[14] Despite such praise for government benevolence, however, the earthquake caused more than financial damage to the bakufu. Although difficult to quantify, the shaking of Edo served as a catalyst that eroded the bakufu's image as the most powerful military organization in the realm.

FIGURE 9 Amaterasu comes to Edo, bestowing wealth as "Tentō" ("the sun"), while an angry crowd attacks the Kashima deity and the Foundation Stone. Courtesy of the National Diet Library, Japan (http://dl.ndl.go.jp/info:ndljp/pid/1304003).

Cowardly Warriors

With the benefit of hindsight, we can see that the Ansei Edo earthquake helped condition the townspeople of Edo—and to some extent a broader segment of Japan's population—for the bakufu's collapse approximately twelve years later. There is a telling entry in *Fijiokaya Diary* dated 1867/9/27 that records a popular song in the context of severe reductions in stipends for bakufu retainers. It states that the thirteenth anniversary of the great earthquake is approaching, an event that half destroyed the military houses. Using plays on words such as the possible meanings "self" and "earthquake" for *jishin* (written in kana script), the second half of the song says that owing to their inactivity, the bannermen have now managed to half destroy themselves on their own.[15] Although these lines of verse focus on the severe financial problems of the bakufu and its retainers, the earthquake was also a blow to bakufu military prestige, particularly because of its timing.

In 1793, in the context of putting off Russian requests for treaty negotiations, the bakufu under the leadership of Matsudaira Sadanobu declared Japan off-limits to Westerners from any country other than Holland. Prior to this time, only ships from Spain and Portugal were formally prohibited from traveling to Japan. This 1793 declaration was framed as long-standing Tokugawa policy, but it was actually a creation of the moment. It helped Sadanobu avoid the difficulties of treaty negotiation, but it put the bakufu in a potentially precarious situation in terms of its domestic legitimacy. Owing to a combination of factors such as the weakening of Dutch naval power in East Asia, the flourishing of trade with China, and whaling in the Pacific by vessels from the United States, non-Dutch foreign ships began appearing frequently along Japan's coast by the early nineteenth century. In 1825, bakufu officials realized the potential damage of Sadanobu's 1793 rule by trying to enforce it. For the first time, the shogunate declared that all prohibited foreign ships would be driven away from Japan's shores, by force if necessary. Not only was the policy impractical, it set up a clear benchmark for assessing bakufu legitimacy.[16]

The arrival of Matthew Perry in 1853 and his subsequent visit and negotiation of the Treaty of Kanagawa in 1854 became an indication to many Japanese that the bakufu had failed a crucial test of its legitimacy. It was not that Japanese elites or ordinary people were especially xenophobic. Views about Japan's proper relations with foreign countries in general and specific Western countries in particular varied across a wide range. Nevertheless, at a minimum, Perry's visits demonstrated bakufu weakness. One concrete manifestation of this weakness was the rushed construction of the offshore artillery batteries, which had a dramatic impact on the local economy and served as a visual reminder of the bakufu's primary function as a military organization. That the earthquake destroyed many *fudai* daimyō residences and several key bakufu offices was a bad sign from the cosmic forces, and the destruction of the artillery batteries was perhaps the most dramatic indication of bakufu military vulnerability.

This weakness was apparent to Edo's townspeople. An excellent example is *Tumultuous Catfish Taiheiki* (*Namazu Taiheiki konzatsubanashi*), by Kanagaki Robun, writing as Daidō Sanjin. A parody of *Taiheiki*, a classic military epic, *Tumultuous Catfish Taiheiki* took the form of an armed struggle between Kashima and the earthquake catfish and his supporters. The sudden night attack of Catfish Slimealot (Namazu Nurakurō) and his allies such as Jumping Fire was devastating. The attackers faced the "cowardly warriors" (*okubyō musha*) of Edo, whose names included Shriveled Body (Karada Chijimaru) and Protruding Eyeballs (Menotama Detarō). The cowardly warriors fled, causing Amaterasu to convene the assembled deities and appoint Kashima Daimyōjin as commander and Atago Gongen as vice-commander to lead the *kami* and buddhas in subduing the rebellion. In the struggle, the forces of order prevailed and the slain catfish were sold to restaurants specializing in broiled eel. The tale ends with the reassurance that Commander Kashima, General Atago, and the myriad deities would return to their home shrines (from Izumo), "calm the waves of the four seas, and ensure, unchanged for myriad generations, the blessings of an unshakable reign (*miyo*)."[17] The emphasis on the *kami* and buddhas, each with a specific realm to oversee and as a group ensuring the prosperity of Japan, was the essence of the popular conception at the time of Japan as *shinkoku*, a country of deities.

Perry's visits were the first dramatic indication of limits to bakufu power, and the earthquake continued this process. Consider alternate attendance. As Constantine Nomikos Vaporis points out,

Alternate attendance was a bedrock of the Tokugawa polity. Despite the weakening of Tokugawa power that scholars have noted in other areas, it was one of the most durable of the shogun's political controls over the daimyō. In institutional terms it underwent little change. Repeatedly issued in the "Laws for Military Houses" (*Buke shohatto*), the attendance requirements were altered only once from their formal inception in 1632 until 1862. In this regard alternate attendance stands perhaps as the single exception to the general relaxation of control measures by the shogunate over the course of the Tokugawa period.[18]

Just as the earthquake shook the bedrock under the city of Edo, it shook the "bedrock of Tokugawa polity" insofar as the bakufu temporarily relaxed attendance duties soon after it struck. The daimyō most adversely affected by the earthquake were precisely those on whom the bakufu most relied for the coastal defense of approaches to Edo and other important duties. Not only did the earthquake render many of these daimyō allies temporarily ineffective, but also they became a major financial drain on the bakufu. After 1855, there was no further attempt to construct artillery batteries to defend Edo Castle. The earthquake financially weakened the bakufu and its major allied daimyō relative to the *tozama* (outside) daimyō, who had less at stake in the existing *bakuhan* state. Both financially and psychologically, the earthquake helped erode bakufu power.

Anxiety in the Divine Land

Perry and the Earthquake

Matthew Perry's fleet arrived in July 1853 and stayed for slightly over one month. He returned with a larger fleet in February 1854 and signed the Treaty of Kanagawa at the end of March. When word of the initial visit reached the imperial court, it sponsored a series of formal prayers at Ise and the other Twenty-Two Shrines that enjoyed at least nominal imperial court sponsorship. The final prayer, offered at all the shrines, called on the deities to keep the "land of deities" free from pollution and harm, to bring peace to the people and the national essence (*kokutai*), and to keep the realm prosperous and militarily strong.[19] The townspeople of Edo became aware of this attempt to mobilize the cosmic forces, and not all of them regarded it as useful. One comic verse that circulated in Edo read, "The divine wind was far in the past! / Some flattery for the *kami* and buddhas of ancient times."[20] The verse points out the absurdity of the imperial court relying on ancient practices to deal with contemporary problems. Even if court prayers conjured up a divine wind (*kamikaze*) at the time of the Mongol incursions during the thirteenth century, some last-minute flattery of the deities would hardly be effective now.

One favorite theme of comic verses after Perry's arrival was to use plays on words in connection with de facto bakufu head Abe Masahiro, more commonly known by his ceremonial title, Lord of Ise (Ise no kami), a homonym for "deity of Ise" (Amaterasu). "Abe" also fit into the word *abekobe,* meaning opposite or upside-down. One example is, "Quite the opposite (*abekobe*) of the Mongol incursions of old / The divine wind of Ise kicks up no wind or waves," with the use of Chinese characters and kana script optimized to bring out the double meaning of "Lord of Ise" and "Kami-wind of Ise."[21] Verses like this one, while poking fun at the historical irony of the peaceful, welcoming stance adopted by the Lord (*kami*) of Ise in the face of a foreign military fleet, do not express a clear political stance. Some did. For example, "Because there is no divine wind from Ise these days / May the former Mito-sama arise."[22] This verse advocates the policy of expulsion and expresses hope that the deposed daimyō Mito Nariaki's voice will again influence bakufu leaders. It was known at this time that bakufu officials were consulting with Nariaki. Explicitly or implicitly informing these comic verses and many others like them was the notion of Japan as a *shinkoku,* a country of *kami* and buddhas.

Popular views regarding Perry and foreign policy ranged from chauvinistic advocacy of expulsion to welcoming a broadening of foreign contacts. Most townspeople, however, did not hold strong opinions either way. They often observed events with a mixture of curiosity and mild apprehension. The apprehension came from the possibility that warfare might break out. This possibility was also a source of profits, as one comic verse, a parody of Emperor Tenji's poem that begins the *Hyakunin isshu* (Hundred people, hundred verses) collection, points out: "The merchant houses sell military and equestrian goods / The money of which we are so fond rains down."[23] Some comic verses expressed a sense of bemused observation. For example, "Paying no heed to the curse [*tatari*] of the divine wind of Ise / America attaches itself to *abekobe*," with the last word meaning both an upside-down situation and Abe Masahiro.[24]

Looking more closely at popular rhetoric in connection with Perry's visits, we find a sudden increase in terms and concepts loosely connected with the idea of Japan as *shinkoku.* The emperor personally did not appear in this discourse, almost certainly because he was an abstraction. However, the imperial court, the Ise Shrines, divine wind, and the Mongol incursions were on the lips of many townspeople. Several comic verses, for example, took the form of imperial proclamations.[25] The imperial court, bakufu, and the American sailors all became grist for the mill of irreverent parody. A few decades later, such comic treatment of the imperial court would become a crime, but in the Bakumatsu era of the 1850s and 1860s, the imperial institution possessed nowhere near the awesome social stature it would acquire in the modern era.[26]

Some observers in Edo linked Perry's arrival with recent events. In the summer of 1853, most parts of Japan were experiencing a drought severe enough to produce state-sponsored prayers for rain at major shrines and temples. One comic verse posits a causal link between the "hairy foreigners" and a "mountain of problems," and it features the word *amerika,* which could be read both as "America" and as "rainfall." With Perry's visits having made such a strong impact on all levels of society, it is not surprising that a major earthquake shaking Edo about a year and a half after the Treaty of Kanagawa would seem anything but random. While the precise meaning of the earthquake might be hard to decipher, many denizens of Edo assumed it must somehow be connected with Perry's visits and perhaps even other events.

One interesting feature of the recollection and listing of past earthquakes in 1855 is that the vast majority of accounts focused only on Japan. Of the major earthquake literature, only *Ansei Chronicle (Ansei kenmonshi)* takes up the previously common rhetorical technique of invoking China, introducing the three-volume work by explaining that even the ancient sage kings Yao, Shun, and Tang (first Shang king) faced natural disasters.[27] *Ansei Chronicle* does not dwell at any length on China, nor does it mention specific earthquakes. In Edo of 1855, China was an abstraction, a land far across the sea at the outer limits of the imagination of most readers. Matthew Perry and his "black ships" eclipsed China as the foreign land in the forefront of the popular imagination. The black ships became connected with the earthquake as part of a broader process of imbuing the recent past with new meaning. Indeed, *Fujiokaya Diary* reports the point of view that because Edo has never experienced such a severe earthquake, the shaking is surely connected with the arrival of the American fleet and presages warfare in the future.[28] Perry and his black ships make several appearances in the catfish prints, sometimes as a relatively sinister presence, sometimes as a desirable presence.

Tension between Japan and the United States is evident in a print featuring Matthew Perry and the earthquake catfish in the midst of a neck-to-neck tug-of-war.[29] Although there is no obvious victor, the catfish seems to be getting the upper hand as Perry lurches forward slightly, and the referee, a plasterer, points with his trowel toward the catfish. The two are at political odds, and the catfish begins a lengthy dialogue: "You stupid Americans have been making fun of us Japanese for the past two or three years. You have come and pushed us around too much. . . . Stop this useless talk of trade; we don't need it. We are sick of hearing the noisy calls of the candy sellers. Since we don't need you, hurry up and put your back to us. Fix your rudder and sail away at once."[30] Perry's reply emphasizes an idealized view of his country's political and social organization: "What are you talking about, you stupid catfish! Mine is the country of benevolence and compassion. No matter what a person does, even if he is a laborer or a hunter, if he is benevolent he can become king." The theoretical possibility that commoners in the United States might become heads of state was fascinating to many Japanese at this time. The passage above ends with an admission by Perry of America's sole problem: a lack of sufficient food. It is for this reason that Perry has come to Japan seeking trade.

The response by the catfish reiterates the view of America as a land characterized by its mode of government and contrasts it with Japan's distinctive quality: "Shut up Perry. No matter how often you brag that your federation is a country of benevolence, if you don't have food you must be poor. If America had the Buddha or the gods, then you would have a good harvest of the five grains. But since you don't, you have to depend on piracy and steal your food. Knowing this, the gods of our country have gathered together and have caused a divine wind to blow and sink your ships and those of the Russians. For sure in the eleventh month of last year the gods struck out against your rudeness." Here we see yet another clear articulation of Japan as a land characterized by the presence of benevolent deities, who provide bountiful harvests—typical rhetoric we have seen in connection with other early modern earthquakes. Moreover, the print regards both the Ansei Tōkai and Ansei Edo earthquakes as attempts by these deities to shake off the foreign presence. Abe Yasunari points out that just as some catfish prints characterized the Kashima deity as inadequate to deal with the current crisis and thus in need of augmentation from outside deities, the view in this print is similar. The only force that might balance the power of Perry and the new foreigners was the collective body of the deities of Japan. Significantly, the dialogue has effectively extended the earthquake in Edo to encompass Japan as a whole.[31]

The dialogue continues with Perry invoking the American spirit: "You catfish! It is funny for you to speak like that, making up your own reasoning.

Despite the fact that men can usually hold you down with a gourd, on the fourth day of the eleventh month you tried to send us away by shaking Shimazu and Numazu, but our American spirit remained unmoved." The printmaker portrays Perry as aware of the Ansei Tōkai earthquake, and the earlier earthquake here becomes an integral part of a larger process of change. It is also significant that there is no mention in this dialogue of government in Japan. The implied contrast is between an aggressive America with an effective government and a Japan whose government is not necessarily effective and whose people must therefore turn to the deities. This dialogue adumbrated both the idea of Japan as an imagined community extending beyond Edo and the reliance of this community on divine intervention.

The print, however, stops well short of advocating any further shakeup of society. The plasterer referee gets in the last words: "Both of you be quiet . . . look with your eyes and see the cracks in the warehouses. We are asked to patch up these cracks and holes, asked over and over again; we are asked to prop up the broken-down walls; we are known for our fine work with the trowel. Everyone admires our work. We are thankful this time for the earthquake, but both of you try to resolve your differences without causing us any more trouble. We don't want to see it; stop it!" The plasterer's view here was probably typical of many townspeople. For the most part, they were happy with the immediate post-earthquake situation, although the process of arriving at that point had been tumultuous, terrifying, and, for some, deadly. So, while thankful for the recent earthquake, many towns-people hoped for an end to major upheavals in the near future. We know from hindsight, of course, that upheavals were just beginning.

Just as the threat of war from Perry's arrival was an opportunity for some townspeople to profit, so too was the earthquake. Several catfish prints feature a deity of good fortune, Daikoku (or Daikokuten), who bestows wealth by means of a magic mallet.[32] Daikoku's name means "Big Black," and, of course, there was another "big black" on the minds of many of Edo's residents—the black ships. The giant earthquake catfish in the 1855 prints were also black. Daikoku the deity, therefore, connected two other "big blacks": Perry's steamships and the earthquake catfish. In the print *Giant Catfish Shaking Great Edo* (*Ōnamazu Edo no furui*), a giant catfish appears to have partially morphed into a whale (fig. 10). It is spouting money, but not from where a whale's blowhole would be. Instead, the coins spout from the location of the smokestack on steamships, making the whale-catfish

FIGURE 10 *Giant Catfish Shaking Great Edo (Ōnamazu Edo no furui,*
1855). Courtesy of the National Diet Library, Japan (http://dl.ndl.go.jp/
info:ndljp/pid/1302035).

resemble one of the black ships. Moreover, the text of an accompanying song includes a play on words that links the homonyms "great country" (*daikoku*) with "Big Black" (Daikoku). Standing on shore, people beckon the whale-catfish-steamship to come closer. This print appears to portray Perry's visits and the trade likely to result therefrom in a generally favorable light.[33]

Abe Yasunari, however, points out an ambiguity concerning the black ships in this print. The short song reads, "The earth of the great country moves, piling up a mountain of treasure in the midst of the city." "The earth of the great country" is *daikoku no tsuchi,* which is a homonym for "the mallet of Daikoku." Why, Abe asks, have interpreters of catfish prints not considered taking the song at its face value, with Japan as the "great country"? The song is written in cursive script, angled into the spout of coins coming out of the creature, and it is written upside down. The only way to read the song easily is to turn the print upside down. Doing so reveals a different landscape. What was originally the oddly red sky looks like earth in the new perspective, and the spout of coins becomes a mountain of treasure. Furthermore, the whole scene now seems focused on "this" shore—that is, on Japan. Abe does not insist that the upside-down reading is the only correct one but rather that it reveals an additional possibility: that Japan might be or become wealthy and great. Here, the term "great country" potentially refers both to the United States and to an optimistic vision of Japan's future, each linked by the figure of Daikoku, the deity of wealth.[34]

Varieties of *Shinkoku*

Whether in 1855 or earlier, major earthquakes were sufficiently traumatic to prompt reassuring talk of Japan as a *shinkoku* of great vintage. This term, usually translated as "divine country," became a key component of nationalist ideology and rhetoric in the modern era. Therefore, it is essential to consider the origins and major changes in meaning of the term *shinkoku* over the centuries, with particular attention to the possible meanings of the term during the early and middle nineteenth century. The first extant appearance of *shinkoku* was in the *Chronicles of Japan* (*Nihongi* or *Nihonshoki*) in the context of Empress Jingū's military campaigns in the Korean peninsula. Until approximately the eleventh century, the conception of *shinkoku* was closely connected with the veneration of the deities of heaven and earth (*jingi*) by the imperial court. The basic idea was that these deities would preserve the imperial line. The late seventh-century move by sovereigns to style themselves "emperor" (*tennō*) instead of "great lord" (*ōkimi*) was part of an attempt to place themselves categorically above the other noble clans. One result of this move was the merging of the various clan deities, which had hitherto been independent, into a hierarchy with the solar *kami* Amaterasu at the top. Therefore, the basic early meaning of *shinkoku* was imperial veneration the *kami,* of whom Amaterasu ranked highest and protected the imperial court.[35]

Beginning in the late eleventh century, corresponding approximately to the rise to power of cloistered emperors, *shinkoku* became part of Buddhism. The basic idea was that Japan was a remote land (*hendo*) in a degenerate age (*mappō*). In this context, *shinkoku* often meant a "barbarian" or an "underdeveloped" country. To take a twelfth-century example, "Our country is a *shinkoku* that has not yet heard Buddhist teachings. It is a land in which day and night people engage in hunting and fishing. What can enlighten it?"[36] There being otherwise no hope for such a place, the compassionate buddhas and bodhisattvas on the "other shore" (*higan*) manifested themselves in Japan as *kami*. These *kami* used rewards and punishments to make people's behavior conform to basic Buddhist norms and guide them toward enlightenment. The idea of *shinkoku* became inextricably linked with the Buddhist concept of foundation and manifest traces (*honji-suijaku*). In this view, Amaterasu became a manifestation of Mahavairocana (Dainichi), the solar Buddha. Although the mainstream tendency was to regard Japan's status as a *shinkoku* as a sign that the country needed extraordinary assistance, especially after the Mongol incursions, a countervailing interpretation within the same framework held that Japan was actually superior to other lands because it alone enjoyed the benefits of thousands of manifest *kami*.[37]

Kitai Toshio argues that the Warring States era, roughly 1470–1580, was a major turning point in *shinkoku* discourse. His basic argument is that Yoshida Kanetomo's (1435–1511) Yūitsu Shintō (One and only Shintō) turned older ideas on their head. Yoshida claimed that Shintō in Japan gave birth to Confucianism and Buddhism in China and India, and that Shintō was the root, Confucianism the branches, and Buddhism the flowers on those branches. According to a Yūitsu text, "Our country is a *shinkoku*. Its way is Shintō. Its ruler is the divine emperor. Its ancestor is Amaterasu Ōmikami." Moreover, according to Kanetomo, all of the residents of the Japanese islands are united as descendants of the *kami*.[38] Precisely at the low point of actual imperial power and prestige, the emperor became a deity in the realm of rhetoric. For obvious reasons, many members of the imperial court embraced Kanetomo's teachings, which spread via daimyō of the Warring States era. Indeed, this same logic lay behind the considerable resources Oda Nobunaga, Toyotomi Hideyoshi, and Tokugawa Ieyasu spent to deify themselves. Many of these Yūitsu Shintō ideas carried over into early modern *shinkoku* discourse.[39] Satō Hiroo points out that in the middle ages, *shinkoku* referred to Japan's peculiar circumstances, whereas in early modern discourse it was often a statement of Japanese superiority. Moreover, early modern *shinkoku* discourse lost sight of the "other shore"— that is, the world of foundation (*honji*) buddhas. Therefore, the *kami* and buddhas in early modern Japan tended to exist independently of any universal principles or cosmology.

Shinkoku discourse in early modern Japan is often associated with nativism (*kokugaku*), one of the schools of thought modern scholars often apply to the Tokugawa period. As Anna Beerens has demonstrated in her innovative study of mid-Tokugawa intellectual networks, however, there was extensive interaction and association between people supposedly located in opposing intellectual camps.[40] *Shinkoku*, whether stated explicitly or in terms of ideas associated with it, was widespread throughout academic and popular discourse, as we have seen in examples from Perry's visit and from earthquake rhetoric. The frequency with which the term *shinkoku* and its associated ideas appear in discourse can serve as a rough barometer of the times. Increased frequency indicates widespread anxiety or a sense of crisis, not the influence of a particular academic school. With this caveat in mind, a very brief look at the *shinkoku* ideas of two scholars often associated with nativism is helpful for our understanding of the term and in contrasting academic conceptions of it with popular discourse.

One theme that emerged in *shinkoku* theory during the Warring States era is that although Japan is a small country, it is culturally superior to other lands. Motoori Norinaga (1730–1801) is famous for advancing a version of this argument, and he stated five reasons: "(1) Being the land where the Sun Goddess [Amaterasu] was born, Japan was the fountainhead of all other nations. (2) Japan's imperial line was unbroken since the beginning of time. (3) Japan possessed the only classics containing the gods' true revelations. (4) Japan produced the world's best rice. (5) Japan had never been conquered by foreign powers."[41]

Turning to popular discourse in the examples we have seen, Amaterasu was well known, either by that name or by synonyms such as Ise(-sama) or Tentō(-sama). However, the claim of Japan as a fountainhead of other nations was largely absent. Popular discourse reflected a vague sense of Japan's imperial line reaching far back into the past, but the specific claim of its being unbroken was not widespread outside of academic circles until the modern era. Norinaga's third reason was also limited to the realm of academic discourse. Variations of reasons four and five, however, appeared commonly in popular conceptions of Japan in the early nineteenth century. The divine wind that destroyed the Mongol fleet was well known, as was the notion that the major benefit of Japan's myriad deities was abundant harvests of grain.

Aizawa Seishisai (1782–1863) was a retainer in the Mito domain, a center for antiforeign sentiment. His *New Theses* (*Shinron*, 1825), written in classical Chinese, circulated privately from 1829, more widely in Japanese language editions during the 1850s, and was finally published in 1858. Many of Seishisai's ideas about Japan and its superiority overlap with Norinaga's five reasons. Seishisai, for example, argued that Amaterasu bestowed rice upon the people of Japan out of concern for their welfare.[42] This idea reverberated in popular discourse, as we have seen, albeit in the more general form of the *kami* and buddhas ensuring bountiful harvests of rice and other grains. Seishisai's opening sentences of *Shinron* reflect many of the Warring States era notions of Japan as a *shinkoku,* plus a geographic dimension to Japanese superiority: "Our Divine Realm is where the sun emerges. It is the source of the primordial vital force . . . sustaining all life and order. Our Emperors, descendants of the Sun Goddess, Amaterasu, have acceded to the Imperial Throne in each and every generation, a unique fact that will never change. Our Divine Realm rightly constitutes the head and shoulders of the world and controls all nations."

Seishisai provided his own gloss on these statements: "The earth lies amid the heavenly firmament, is round in shape, and has no edges. All things exist as nature dictates. Thus, our Divine Realm is at the top of the world. Though not a very large country, it reigns over the Four Quarters because its Imperial Line has never known dynastic change. The Western barbarians represent the thighs, legs, and feet of the universe. This is why they sail hither and yon, indifferent to the distances involved. Moreover, the country they call America is located at the rear end of the world, so its inhabitants are stupid and incompetent."[43]

In the late 1850s, Seishisai's worldview became a rallying cry for advocates of *sonnō-jōi* (revere the sovereign, expel the barbarians), a term that served more as an anti-bakufu slogan than as a description of a realistic policy choice. By contrast, popular *shinkoku* discourse of 1855 and ear-lier often implied Japanese superiority, but not in such strong terms. The chauvinistic antiforeignism of Seishisai and the activist samurai he inspired was often absent or muted in popular discourse. From their standpoint as observers of the events of the day, many of Edo's townspeople viewed the foreigners and their ships more with curiosity than with fear or disdain.

Popular commentary on Perry's visits had no influence on bakufu pol-icy. Prints and other popular media linking Perry's visits with the Ansei Edo earthquake likewise had no immediate political significance. Nevertheless, the earthquake was a catalyst for the creation of broader webs of associa-tion, some of which did not bode well for the bakufu as it faced increas-ing challenges. Anxiety in the divine land helped weaken the *bakuhan* state from the bottom up.

Amaterasu Comes to Town

Perhaps the most significant role of the Ansei Edo earthquake in shaping popular perceptions within the border framework of Japan as a land of dei-ties was in bringing Amaterasu to town. We have already seen one example in the form of the anti-Kashima print celebrating the arrival of Tentō. It may seem odd to say that Amaterasu was not in Edo prior to late 1855, and I do not mean to suggest that Amaterasu was unknown. Clearly, as we have seen, there was a widespread, if vague, knowledge among Edo's townspeople about the deity of Ise. Moreover, the Ise Shrine complex was the focus of popular pilgrimages during many *okage* years. Consider, however, the basic plot of *Namazu Taiheiki konzatsubanashi*. Amaterasu resided in Ise and presided over meetings of the major deities at Izumo during the tenth lunar month. There he appointed Kashima and Atago to go back to Edo to deal with the catfish rebellion. In other words, Amaterasu remained far from Edo. Notice also my choice of pronoun. One major dif-ference between academic discourse and popular discourse, at least in Edo, was Amaterasu's gender. Amaterasu was a goddess to academics, but in the catfish prints, he appeared either as an abstraction (e.g., *gohei*, shining folded paper) or in male guise. For example, in one talismanic catfish print, *Earthquake Protection (Jishin no mamori)*, Amaterasu is clearly labeled and appears prominently as a man with a moustache. In this print, Amaterasu appears with Kashima and other *kami* of Edo and declares in part that the bearer will receive protection from the deities (*shoshin*) "above all the land of Japan . . . all the way down to the Gold Layer [*konrin*], the abode of King

Yama (Enma-ō)." In the lengthy text of the anti-earthquake prayer in the print, Kashima implores the myriad deities to "make the five grains thrive" and later that they ensure peace within the realm, an abundant harvest, and protect the ruler (*kimi*). The scope of the protective verse is clearly Japan, not simply Edo.[44]

Interpreting the earthquake as an intervention for world renewal (*yonaoshi* as wealth redistribution) by the cosmic forces was not without some difficulties. For one thing, not all occupations benefited, as we have seen. More problematic was the death of innocent people, such as the filial daughter in the *Ansei Record* (*Ansei kenmonroku*) tale. The metaphor of powerful medicine with strong side effects undoubtedly assisted in permitting the world renewal interpretation. Particularly after the number of deaths became widely known and it was clear that rumors had exaggerated the toll, it may have been easier to regard innocent deaths as something like what today we would call collateral damage. It would also have been easier for townspeople to regard the overall event as beneficial if the higher powers had intervened to try to minimize the loss of innocent lives. Fortunately, they did. Amaterasu sent a divine white horse (*jinme*) from Ise to appear in the sky and shed lifesaving hairs to protect those below.[45] One catfish print features the divine horse charging ahead, strands of hair flying, knocking down the earthquake catfish and nearly running over a hapless Kashima deity. The text explains that the horse appeared in the sky shedding hair, and the people were saved by the power of the Great Shrine at Ise.[46] A similar print explains that those who escaped unharmed by means of the falling hair did so because of their faith in the Great Shrine at Ise and the various *kami*.[47] The deities that in ancient times protected the imperial court now protected ordinary people throughout *shinkoku* Japan.

Although the Ise Shrine complex might be able to extend protection to any place in Japan, Amaterasu was still only the most powerful of the thousands of *kami*. He ("she" in academic circles) was closely associated with the territory of Ise Province, which was even known as "Shinkoku" in some late medieval texts, with "-koku" indicating a province.[48] Despite the prevalence of *shinkoku* discourse among Edo's townspeople, however, it is important to note that Amaterasu remained at Ise until the time of the earthquake. Prior to the Ansei Edo earthquake, Amaterasu had little or no direct connection with Edo in the popular imagination. There was no formal cult of Amaterasu worship in Edo, and Kashima Daimyōjin was the dominant deity of the region. Just as there was no clear central government in 1855, so it was with the deities. Kashima was largely autonomous, but strictly speaking, Amaterasu outranked him. This difference in rank came into play only after Kashima (or Ebisu) failed to prevent the earthquake. Similarly, in academic circles, scholars such as Aizawa Seishisai regarded the bakufu as exercising authority that ultimately derived from the imperial court. The bakufu's foremost task was to protect the realm against foreign incursion, much like the Kamakura bakufu had done in the face of Mongol threats. In many eyes, the Tokugawa bakufu failed in 1853 and 1854. Moreover, the bakufu looked weak again when the earthquake destroyed its artillery batteries and major offices. In the cosmic realm, the earthquake was a clear indication of Kashima's failure to protect his domain, even if his being out of town for a meeting mitigated the blame.

Some early prints convey an irreverent sense of disgust that Edo's local deities would have so badly mismanaged the balance of cosmic forces resulting in the earthquake. In *Catfish and the Foundation Stone* (*Namazu to kanameishi*), fires rage and the earth shakes above the sinister-looking figure of a catfish. Ebisu, filling in for Kashima, looks tired as he dozes against the Foundation Stone.[49] A strange-looking man to the left of the print is the thunder deity. He engages in a peculiar pastime of some townspeople, which we might call "extreme farting" or perhaps "thunder farting." The basic object of this sport was to make more noise than one's opponents. According to the scholar Hiraga Gennai (1729–1779) in his treatise *On Farting* (*Hōhiron*), thunder farting made its debut in 1774 at the Ryōgoku Bridge, a major site of freak shows (*misemono*) and other popular culture performances in Edo. Small drums issue forth from the thunder deity's posterior, no doubt to emphasize the booming sonic element in his performance. The small figure of a hapless-looking man on horseback in the print is Kashima, rushing back from his meeting with other deities in Izumo. These incompetent deities have allowed a major disaster to unfold in the form of fire-ravaged, post-earthquake Edo.[50] Large gold coins fall from the burning city, presaging the world renewal theme that became prominent in later prints.

This disgust at the Kashima deity's ineptitude sometimes manifested itself in a different manner: demoting Kashima and replacing him with Amaterasu. In other words, it was necessary for Amaterasu to come to Edo to restore order to the realm. To some extent, this demotion followed the money in real life. Five days after the earthquake struck, the bakufu paid thirteen shrines throughout the country to conduct prayers. The inner and outer shrines at Ise received five gold bars, and the Kashima Shrine, Katori Shrine, and the others each received three. Bukufu gold thus acknowledged that Ise was Japan's leading shrine and Amaterasu its leading deity. As the focus of *okage* pilgrimages and for other reasons, the Ise shrines had long cultivated an image as Japan's most prestigious religious institution.[51]

Other prints depict Kashima in supporting roles, subordinate to Amaterasu, or in a mildly antagonistic relationship with Amaterasu. In one image, Amaterasu, Kashima, and Hachiman ride horses across the sky of a devastated Edo. Amaterasu orders the earthquake catfish to depart from Edo quickly and dispenses strands of horsehair, his back turned to Kashima. Kashima comments on the severity of the destruction and holds the Foundation Stone aloft, but the stone is of no use in the sky.[52] In this image, Amaterasu and Kashima are of roughly the same size and occupy the same height in the sky. In a different print, however, Amaterasu, called the "imperial ancestor of great Japan," towers above the smaller figure of Kashima, who assists in distributing divine horsehair alongside of six other local Edo deities.[53] Here, Kashima clearly has been demoted. Its text refers to Japan as a *shinkoku* whose people are fortunate that Amaterasu, the emperor (*mikado*), the shogun, and the domain lords are all benevolent and concerned for the people's well-being.[54] Here we see a vision of Japan that includes land, deities, and rulers arrayed in a manner similar to that which the newly founded Meiji state began promoting slightly over a decade later. While relatively few prints are as explicit as this one in positing Amaterasu's superiority over the deities of Edo, many reveal a degree of tension between Kashima and Amaterasu. Prints featuring Amaterasu also suggest that the impact of the earthquake extended beyond Edo to other areas of Japan.

"Japan" appears frequently in the text of the catfish prints, usually as Nihon/Nippon or some variation such as Dai-Nihon. Moreover, the text of catfish prints consistently depicts Amaterasu in terms such as lord of "the skies above all of the land of Great Japan," albeit often in concert with the "various other deities."[55] It is possible, as Takashi Fujitani has argued, that in some rural areas, consciousness of Japan or an awareness of emperors and their principal deity were either absent or so vague as to be no different from local folk beliefs. As a generalization, it was indeed the case that "during the Tokugawa period, Japan was populated by a people separated from one another regionally, with strong local rather than national ties." Moreover, insofar as pre–Meiji Restoration residents of Edo were cognizant of the emperor's existence, they tended to see him as a popular, wish-granting deity.[56] Information networks, however, had long tied urban areas together, and by 1855 townspeople were well aware that they lived in Japan, even if they had only recently begun to pay attention to events outside of their cities or regions. "Japan" as an imagined community in 1855 did not possess as rich an array of cultural attributes as it would acquire in modern times. The Ansei Edo earthquake reinforced the idea of Amaterasu as ruler of Great Japan to an audience who until 1853 or 1854 had little opportunity to think about the solar deity. Although Amaterasu was part of a vast array of characters that appeared on the post-earthquake stage, it is significant that the emperor's deity eclipsed Kashima, closely associated with Edo and the bakufu. It was a cosmic dress rehearsal for the earthly events of 1867 and 1868 that took place at the time of the next *okage* year. Gregory Clancey points out that the 1891 Nōbi earthquake functioned as "a dress rehearsal" for the major "nationalizing" event of the era, the First Sino-Japanese War.[57] The Ansei Edo earthquake played a similar role vis-à-vis the Meiji Restoration. It helped make what would soon happen easier to imagine.

Dancing

The catfish print *Kashima Fear* (*Kashima osore*) portrays seven ecstatic dancers, six townspeople and Enma no ko (child of Yama) surrounding a giant catfish dressed as a representative of the Kashima Shrine.[58] These representatives would travel throughout Japan in the early spring at the start of the New Year, pronouncing oracles regarding the prosperity of the upcoming year. Religious dancing connected with this process, dancing in anticipation of a fruitful harvest, was called Kashima odori or Kashima kotofure. The seven dancers substitute for the seven deities of good fortune. The catfish holds a pole with a red solar disk at its top, and a rabbit appears in the disk because 1855 was the year of the rabbit.[59] The dancers also convey a sense of the carnival-like atmosphere that was part of pilgrimages to the Ise Shrine during *okage* years. The text of the print abounds with plays on words and double meanings, and it explains that despite the destruction the shaking has caused, the people have received the blessings of the ruler (*kimi*, a vague term). Moreover, the Foundation Stone is a sign that the present regime will endure. Then the catfish decrees a propitious oracle for "this *yonaoshi* earthquake." The carpenter mentions his high wages, and Enma no ko exclaims, "*Yonaoshi!* Everything in this world is good, good, good!" The print is a typical example of the interpretation of the earthquake as world renewal.

The appearance of Enma no ko in this print and elsewhere in connection with the earthquake deserves some attention.[60] The appearance of a peculiar person or creature with at least some supernatural capabilities was a common motif in popular riots or other demonstrations that involved mass protest and property destruction, usually in the name of world renewal. The riots in Edo of 1787, for example, lived on in folk memory and popular literature up to the time of the riots there in 1866. As commentators recreated narratives of the 1787 riots over the years, they connected their outbreak to the appearance of figures such as "an unshaven youth of seventeen or eighteen and a man of unusual strength," a *tengu* (legendary bird/man goblins with magical powers), a priest of superhuman strength, "a tiny youth with the strength of a sumo wrestler," and incarnations of popular heroes from the past such as Benkei or Yoshitsune.[61] Enma no ko possessed precisely the qualities of a tiny youth with great strength. His appearance in 1855 fits perfectly a well-established pattern of popular world renewal uprisings. The main difference, of course, is that in this case the shaking earth performed the destruction and redistribution of wealth that would ordinarily have been the work of enraged townspeople.

The other element in this print worthy of closer examination is the frenzied dancing. Dancing was an integral part of folk religion and the mass pilgrimages to Ise during *okage* years. The most famous and significant frenzied dancing took place during the *okage* year of 1867, twelve years after the earthquake. During that year, frenzied dancing occurred widely throughout Japan, known retrospectively as *ee ja nai ka* (approximately, "What the hell?!"), after one of the common refrains the dancers chanted. *Ee ja nai ka* dancing was much like mass pilgrimages except that it occurred locally, usually but not always prompted by reports of amulets or other objects falling from the sky in connection with popular local cults. The classical interpretation of the *ee ja nai ka* dancing is that it constituted anti-bakufu protests. Several scholars, however, have rejected such interpretations or added nuance to them. Takagi Shunsuke, for example, argues that *ee ja nai ka* was a manifestation of a longer tradition of folk dancing in connection with world renewal. In other words, the dancing reflected aspirations on the part of peasants or ordinary townspeople to improve their economic lot and should be regarded as a part of "world renewal as a religion" (*yonaoshi shinkō*).[62] Nevertheless, there was a difference between the dancing in 1867 and that of previous years, both in terms of scale and content. The 1867 dancers were well aware of the implications of broader political developments. In the Kyoto, Osaka, and Kobe areas, for example, dancers began to chant "Thanks to Chōshū," the price of rice has dropped, in celebration of the bakufu's calling off its second expedition against Chōshū. Indeed, a specific dance, *Chōshū-odori*, emerged from this display of bakufu weakness.[63] Spontaneous *zannen-san* pilgrimages in the Osaka area broke out in the form of mass visits to the graves of some fallen Chōshū soldiers. The political implications of these visits prompted the magistrate of Osaka to ban them, thus making them all the more popular.[64]

Knowing what happened twelve years later, we can see that the popular interpretations of the Ansei Edo earthquake, while not explicitly political in the manner of *zannen-san* pilgrimages, were in some cases an intermediary stage between traditional world renewal events and the politically charged dancing and pilgrimages in 1867. The almost carnival-like atmosphere depicted in *Kashima Fear* as the cosmic forces upended society for the benefit of the common people provided an unintentional preview of events soon to come. Moreover, although most instances of *yonaoshi* in 1855 referred to the transfer of wealth from rich merchants to other townspeople, a more ominous sense of this term also emerged.

Ominous Associations

An entry in *Fujiokaya Diary* four days after the main shock describes the dream of a Hongō resident named Shōgorō. He made his living as a gardener and his wife sold cooked potatoes. On the night of the sixth day, three men dressed like warriors came to buy potatoes, but instead of paying in copper cash they handed over a gold coin. Suspicious, Shōgorō refused to accept it, which prompted them to hand over a wooden tag. After Shōgorō refused to accept the tag, they became angry, dragged him through several city blocks, and pushed him into a large hole in the earth. There, Shōgorō encountered eight or nine "large men." They turned to him and showed Shōgorō that the tag read, "Great Earthquake in Shinano"—that is, the Zenkōji earthquake of 1847. The men explained that they were on a mission of "world renewal" (*yonaoshi*), to assist good people and slay those who are evil. They told Shōgorō not to worry, but they warned that they would cause an earthquake each day until their work was finished. At that point, Shōgorō awoke from his dream in a temporary shelter with his wife and child.[65] The men of supernatural strength Shōgorō encountered in his dream were typical of the mythical, heroic figures commonly imagined to have precipitated past riots or mass protests. These men, personifications of past earthquakes, revealed to Shōgorō the two possibilities inherent in their rectifying project: wealth in the form of a gold coin, or destruction represented by the most dramatic and deadly earthquake of the recent past, Zenkōji. What distinguishes the vision of *yonaoshi* expressed in this tale was its potential for deadly violence and the threat that such violence would continue into the future.

Revenge of the Earthquake Gang

Although the literal impression conveyed by the compound *yo* (society, the world) plus *naoshi* (renewal, correction, or rectification) might suggest revolutionary goals, typical Tokugawa-era world renewal events were limited in both their objectives and levels of violence. Stephen Vlastos' summary of the 1868 *yonaoshi* uprisings in Aizu would apply to most other such events during the previous century and a half: "There were precise limits to 'world rectification.' It was a revolt by small cultivators and the poorer members of the village community against high-status peasants. They did not directly question their relationship to the state. Their concerns were local and not national, so that the peasants who sacked village officials and convened popular assemblies also appealed to the Meiji government for benevolence."[66] Similarly, as we have seen, most characterizations of the Ansei Edo earthquake as world renewal were based on a local redistribution of wealth, a forcible cracking open of the storehouses of the rich. Human actors carrying out forcible redistributions of wealth or demanding redress of grievances were constrained in their violence, usually limiting their destruction to property they perceived as having been unjustly acquired.[67] Earthquakes, however, represented a serious escalation of world renewal violence because the shaking earth was free to mete out the death penalty.

The print *Catfish Revenge* (*Namazu no adauchi*) features a masterless warrior catfish named Shindōsai (Shaking purification), described as resembling a warrior more than six *shaku* (roughly two meters) in height and having a strange body and face. Shindōsai explains that he and his associates comprise the "earthquake gang" (*jishinban*). The name of the gang is a play on Jishinban, a neighborhood-based townsperson patrol in Edo and Osaka that combined security and fire-fighting duties. This earthquake gang has been shaking up Japan from one end to the other, starting with Zenkōji. The stated reason was that people have been grilling and eating their catfish brethren. He and his gang killed many in Mino, Ōmi, Kyoto, and Osaka, but still people's hearts are not good. The gang has now arrived in Edo, via Yamato. The rogue catfish then goes into greater geographic detail: "We shook up Yamato, Kawachi, Kii, Izumi, Iga, and Ise, then followed the Tōkaidō to Izu and shook up Shimoda [Ansei Tōkai]. We rested there briefly and then returned to the three provinces of Suruga, Tōtomi, and Owari [Ansei Nankai]. Last month we got as far as Yoshiwara, and now we are here. If you do not give in to our demands, we will shake you to death." The dialogue ends ominously by suggesting that when Kashima gets back from Izumo, the earthquake gang will escape to the northern provinces.[68] To many observers in Edo, Zenkōji was the first step in a larger process of cosmically ordained shakeups and change, which they feared might still be in progress. This print does not use the term *yonaoshi,* but the name of the gang leader implies religious purification. The tale in the print, of course, closely resembles that of the gardener Shōgorō's encounter with a similar earthquake gang.

The route this gang traveled, of course, followed recent past earthquakes. The gang's violent nature may also reflect another recent development: increasing levels of violence in the countryside. Although this phenomenon became especially apparent during the 1860s, the process was well under way by 1855. Beginning in 1804, the bakufu began issuing injunctions against commoners practicing the martial arts, but in Musashi near Edo in the 1850s nearly 80 percent of the members of the local fencing school were commoners. While on the one hand attempting to prohibit commoners from acting like samurai, the bakufu increasingly relied on them for police and militia duties. The result was that "the shogunate's claims to monopolize the legitimate use of violence were being undermined from within and without during the final years of the early modern era."[69] The portrayal of the recent pattern of earthquakes as a violent gang of outlaws may well have been a reflection of the rising tide of violence in society at that time. In any case, it raised the menacing possibility that the violent upheavals of the past eight years were part of a larger, ongoing process.

Webs of Associations

The examples of Shōgorō's dream, the earthquake gang, and many other accounts reveal an awareness of recent earthquakes constituting a meaningful pattern. Zenkōji was most prominent by far. Significantly, although the Zenkōji disaster was known in Edo in 1847, its psychological impact at the time was minimal. The disaster produced little more than a tasteless *senryū* verse lauding Zenkōji in Shinano as a place that provides three kinds of funerals: fire, water, and earth.[70] After the tenth month of 1855, however, Zenkōji suddenly loomed large. A catfish print entitled *Edo Catfish and Shinano Catfish* (*Edo namazu to Shinshū namazu*), for example, features a mob attacking two giant menacing catfish, one with "Edo" on its forehead and the other "Shinano."[71] One implication of this sudden interest in recent earthquakes was a realization that society was undergoing profound changes, which residents of Edo abruptly deduced had been in progress since 1847 or possibly 1830.[72]

The Ansei Edo earthquake changed perceptions of recent history. In this connection, one of the *Ansei Chronicle* authors wrote that the elderly possess the most penetrating understanding of society because they have witnessed the major events of the past thirty or so years. He then offered a list of the sixteen most significant recent events. The list starts in 1829 with the first issue of *issugin*, silver coins worth approximately one-sixteenth of a *ryō*. These coins were reissued in 1853 in connection with the construction of the offshore artillery batteries. Five items are about issues of new currency, one explains the Tenpō famine of the 1830s, one describes the Tenpō reforms as eliminating unauthorized theaters, brothels, and the trade guilds, and one entry for 1838 describes a severe rice shortage. The shogunal visit to Nikkō, a great Edo flood in 1846, the opportunity to hunt deer for a fee, severe lightning damage, and the Ansei Edo earthquake each constitute an entry. Visits by U.S. and British ships are listed in 1852, and the visit of a Russian ship to Osaka in 1854 is another entry, along with the opening up of Hokkaido. Not surprisingly, economic matters predominate, but visits by foreign ships are also prominent—all in the context of discussing the 1855 earthquake. The list of significant recent events ends with the author commenting that he is fortunate to have been born in an age of peace and to have been able to see such marvels as foreign people and the building of Western-style steamships, cannon, and other devices. He characterizes the current earthquake as a momentary pain (*ku*) that will lead to greater enjoyment (*raku*).[73]

Whether this optimistic view of the recent past was typical of the majority of townspeople is hard to say, but if the catfish prints and popular literature are reasonably representative, many of Edo's residents shared it. Viewed relatively optimistically, Japan was in the midst of an unprecedented period of change. While this change entailed some terrifying or painful aspects, it was also ushering in a fascinating new age of potential opportunity. The plasterer-referee's words in the catfish print featuring the neck tug-of-war with

Perry are less enthusiastic. In that view, although things have worked out reasonably well until now, he demands an end to unexpected shakeups in the future. Shōgorō's dream and the print featuring the violent earthquake gang represent the dark side of possibilities—namely, punitive death and destruction extending into the future with no promise of benefits. While differing in degrees of optimism, all of these views assumed that the Ansei Edo earthquake was part of a larger process of major change sweeping through Japan, change that was unlikely to end with the shakeup of the shogun's capital.

Linking the earthquake with other recent earthquakes, disasters, and political developments was not limited to townspeople. Warriors thought in similar terms, though they often selected different events to make their case. Bakufu retainers sometimes reflected on the resource drain that recent events had imposed on the shogunate. Consider a letter from bannerman Tsuchiya Kyūba in the seventh month of 1856. It begins by stating that in recent years the arrival of foreign ships has resulted in many expenses and activities. Next, he mentions the fire (in 1854) that destroyed the imperial palace. After that, he remarks on the great destruction caused by the recent earthquake. He also explains that several rivers in the Kantō region have recently flooded, causing extensive damage. Kyūba says he expects that the bakufu will provide assistance in solving these problems but that all the repairs seem like a herculean task, especially with so much earthquake damage to the Edo mansions.[74] Like many throughout society, the earthquake prompted Kyūba to bring together recent events that he regarded as significant or meaningful, in his case with a worrisome outlook.

Conclusions

In the longer term, the Ansei Edo earthquake weakened the bakufu, and it was only the first of a string of natural and social upheavals to hit Edo. First, there was the fierce typhoon ten months after the earthquake. In 1857, a deadly influenza epidemic struck. During the fifth month, the epidemic interfered with government administration in Edo Castle and likely took the life of leading bakufu official Abe Masahiro, who died in the sixth month. The following year, during the seventh and eighth months, cholera struck Edo and surrounding areas. The outbreak claimed approximately thirty thousand lives, roughly three to four times more deaths than occurred from the earthquake. In 1859, a fire destroyed the keep of Edo Castle. Partially overlapping with this string of upheavals was a series of political upheavals: the U.S.–Japan Treaty of Amity and Commerce in 1858, Ii Naosuke's Ansei Purge of 1858–1859, followed by Ii's assassination in the 1860 Sakurada-mon Incident. A line from a satirical song from 1860 summarizing the Ansei era reads literally, "Not calm government [*ansei*], tsunami, earthquake and great storm, cholera, great fire and Sakurada trouble."[75] Soon thereafter, the malevolent forces of nature returned in the form of a severe measles epidemic that swept through most parts of Japan in 1862. In 1863, foreign ships destroyed Chōshū guns at Shimonoseki. Military activities then and in subsequent years pushed the price of rice to record levels. Ordinary people in the capital began to feel a sense of famine as rice became too expensive to eat. Riots occurred in Osaka and Edo in 1866, and the bakufu began its final collapse the following year amid widespread outbreaks of frenzied, carnivalesque dancing.[76]

For most Japanese, regardless of worldview or social position, a major earthquake striking the shogun's capital during an *okage* year was not a random occurrence. Not everyone read the earthquake the same way, but some broad patterns have emerged from the discussion in this chapter. First, because of its proximity in time to Perry's visits and because it destroyed the newly constructed artillery batteries that had become iconic fixtures in popular imagination, the earthquake highlighted bakufu military weakness. Stated in terms of popular literature, the "cowardly warriors" of Edo were unable to subdue the rebellion of Catfish Slimealot without divine assistance. For the actual warriors, the earthquake brought about the first relaxation in alternate attendance, and it drained bakufu finances just as the shogunate was about to face a series of serious challenges.

Second, anxieties connected with both Perry's visits and the earthquake reminded the denizens of Edo that they lived in Japan, which was a *shinkoku*. Comic verse and other forms of popular oral and written discourse adumbrated a view of Japan as characterized by the presence of thousands of benevolent deities. These *kami* and buddhas had long before shed their medieval image as manifestations (*suijaku*) of universal Buddhist principles (*honji*). By the 1850s, they had come roughly to mirror the political situation in Japan, with each major deity overseeing its specific realm. There was an important difference, however. The shogun possessed the greatest military power in the political realm. However, in the divine realm, Amaterasu, although not omnipotent, was clearly the leader of the *kami* and buddhas. Working together, this host of deities ensured that Japan was blessed with bountiful harvests. That they were unable to prevent the earthquake could be explained by the annual convention of the deities at Izumo during the tenth lunar month, thus leaving Edo vulnerable.

Ise was far from Edo, but in the post-earthquake popular imagination, Amaterasu intervened in Edo affairs to bring the situation under control. In some accounts, this intervention was simply Amaterasu charging Kashima and Atago on behalf of the assembled deities to return to Edo and restore order. In other accounts, Amaterasu sent his horse, and in several catfish prints Amaterasu appeared in person to bring both order and prosperity to the distressed city. Therefore, in the months after the earthquake, popular literature outlined what was in effect a dress rehearsal among the deities for what actually happened roughly twelve years later when the emperor came to town to stay. More broadly, even though the immediate focus was usually on the bakufu, earthquake literature frequently made reference to the imperial court as the bedrock of social longevity and stability. One text in the form of a purification chant, for example, points out that for eight thousand generations, the Foundation Stone has supported an unshakable line of sovereigns in the "imperial country of reed plains."[77]

Julia Adeney Thomas has pointed out that one characteristic of the "Tokugawa topographic imagination" was "a particular penchant for discovering a central point."[78] Whether after the Ansei Edo earthquake or other major earthquakes, talk of *shinkoku* brought Japan as a whole to the forefront of attention. Equally important, talk of *shinkoku* led to a central point, and it was not Edo. That point was the imperial court, broadly defined to include not only the emperors in Kyoto but also, more prominently, Ise and Amaterasu. When confronted with the terrifying convulsions of the earth, bakufu power and centrality paled in comparison to a conception of much older social structures. The stress of earthquakes helped popularize a view that Motoori Norinaga earlier articulated: the imperial line was the pivot of the universe.[79]

The theme of redemption or renewal, so common after the earthquake, continued to motivate commoners for the next twelve years. These calls for redemption remained anchored in local issues, but they also became explicitly political and national in scope. As George M. Wilson points out regarding studies of the Meiji Restoration, "What is missing is attention to the much more dynamic manifestations of the popular anxiety that permeated the very atmosphere of the tumultuous *bakumatsu* years. Throughout the country during the mid-1860s, throngs of pilgrims congregated at religious sites, especially the complex of shrines at Ise in central Japan, where the regalia and mystique of the imperial house's mythical past resided." These pilgrimages had occurred since the late sixteenth century, but "what changed in the 1860s was the intensity and the random quality of the pilgrims, whose behavior verged on the hysterical."[80] Especially insofar as it focused awareness on Ise and Amaterasu, the Ansei Edo earthquake prefigured developments of the next twelve years.

The earthquake did not produce active opposition to the bakufu among Edo's townspeople, but it did raise doubts about bakufu strength. Moreover, townspeople's support for the bakufu was economic, not ideological. The bakufu never devoted serious resources to producing an ideological justification for its existence for consumption by common people. When Edo's economy declined after what turned out to be a permanent relaxation of alternative attendance in 1862, the economic incentive for the townspeople to support the shogunate faded. As bakufu power also faded, it was not difficult psychologically for Edo's population to welcome the emperor, who had long enjoyed the reputation of being a benevolent quasi-deity. Looking back from 1868, popular consciousness of the imperial court and its deities increased significantly owing to the events of 1853, 1854, and 1855.

The Ansei Edo earthquake was the first major earthquake characterized as an instance of world renewal. For the most part, this characterization simply substituted the shaking earth for masses of protestors or rioters. However, unlike human agents of world renewal to that point in time, the earthquake was significantly more violent and deadly. Moreover, the earthquake caused many observers to regard other recent upheavals, human or cosmic, as part of a broader pattern of fundamental change. Significantly and sometimes ominously, there was no indication in late 1855 or early 1856 that this pattern of change had run its course. Therefore, I argue that although the Ansei Edo earthquake was not a revolutionary event in and of itself, some of its effects helped condition society to anticipate major change in the near future. The *bakuhan* state emerged from the shaking of 1855 standing but weakened in ways that were not immediately apparent.

CHAPTER 6

Into the Twenty-First Century

As a general model of the effects of earthquakes over time, Jelle Zelinga de Boer and Donald Theodore Sanders propose the metaphor of a vibrating string: "If we think of an earthquake as the plucking of a long, tight-stretched string representing time, the string will vibrate. During the quake itself . . . the vibrations will have high amplitudes and short wavelengths. They will be powerful, but each will last only a moment. Farther along on the string, with the passage of time, the amplitudes will decrease and the wavelengths increase. That is to say, the aftereffects will become less intense and they will last longer."[1] The Ansei Edo earthquake conforms to this pattern. An example of a high-amplitude, short-wavelength effect was the redistribution of wealth so prominently discussed in prints and other popular media. Less than a year after the main shock, the influx of workers from outside Edo had substantially eliminated windfall profits. The political effects, including the strain on bakufu finances, were less obvious but longer lasting. This chapter examines the low-amplitude, long-wavelength side of the spectrum, exploring ways that the Ansei Edo earthquake has influenced modern and contemporary understandings of earthquakes in Japan.

The impact of the Ansei Edo earthquake lasted far beyond the fall of the bakufu. The event persisted in popular memory and as a reference point among Japan's pioneer earthquake scientists and architects. During the late nineteenth and early twentieth centuries, seismologists developed tools and techniques for measuring earthquake-related metrics, but they lacked an understanding of the geophysical mechanisms that cause earthquakes. An appreciation of the significance of faults developed early in the twentieth century, but the theory of plate tectonics did not become widely accepted in

Japan until the 1970s. In Japan, one result of this gap between the capacities for measuring versus explaining was that lore from the Tokugawa period continued to influence modern seismological thinking and investigations. My basic argument in this chapter is that key elements of the Tokugawa past have conditioned modern and contemporary Japan in the realm of popular thought, in the development of seismology, and in perceptions of Japan and its relationship with earthquakes.

A comprehensive discussion of the history of seismology is beyond the scope of this study. However, because Western science and the accumulated body of Japanese earthquake observations and lore merged during the Meiji era, a brief consideration of understandings of earthquakes in the Western world provides useful context. Here I focus on pioneer geologist Charles Lyell (1797–1875), whose major work, *Principles of Geology,* was roughly contemporaneous with *Thoughts on Earthquakes* (*Jishinkō*) in Japan, both temporally and in the extent of its impact.

Writing in the middle of the seventeenth century, René Descartes posited water and water vapor as causing earthquakes, "when exhalations, collected and ignited in the earth's cavities, suddenly rarefied causing the earth to shake."[2] By the time Lyell published vast, detailed information on past earthquakes from many parts of the world, the quantity of geological knowledge had grown enormously compared with Descartes' day. Earthquake causes, however, remained an enigma.

Lyell regarded the effects of earthquakes as instrumental in shaping the natural landscape, but he was unable to offer a convincing theory of their causes. One theory he entertained was that waves in the earth's molten interior caused earthquakes. He also considered electrochemical theories and the possibility that escaping gas, liquefied under pressure, produced the force necessary to shake the earth.[3] Discussing connections between earthquakes and volcanoes, Lyell proposed a mechanism similar to but more complex than the Japanese idea of yang trapped within the earth: "If earthquakes be derived from the expansion by heat of elastic fluids and melted rock, it is perfectly natural that they should terminate, either when a volcanic vent permits a portion of the pent up vapours or lava to escape, or when the earth has been so fissured that the vapour is condensed by its admission into cooler regions, or by its coming in contact with water. Or relief may be obtained when lava and gaseous fluids have, by distending the strata, made more room for themselves, so that the weight of the superincumbent mass is sufficient to repress them."[4]

Similarly, in the context of wavelike motions of the earth's surface during earthquakes, after likening strata in the earth to a large carpet or rug shaken up and down at one end to produce waves, Lyell explains, "In like manner, a large quantity of vapour may be conceived to raise the earth in a wave, as it passes along between the strata which it may easily separate in an horizontal direction, there being little or no cohesion between one stratum and another."[5] Even though he understood earthquakes in a manner roughly similar to many of his contemporaries in Japan, Lyell's knowledge of the broader geophysical context, including the earth's strata, the potential roles of pressure, and wave mechanics exceeded the knowledge available in Japan. In hindsight, of course, it is easy to see what was missing: an understanding of faults and the forces that act on them.

As for faults, Lyell did use the term in discussing the dramatic sinking of the marble quay during the Lisbon earthquake of 1755: "In this case we must either suppose that a certain tract sank down into a subterranean hollow which would cause a 'fault' in the strata to the depth of six hundred feet, or we may infer, as some have done, from the entire disappearance of the substances engulphed, that a chasm opened and closed again."[6] The basic definition of a fault is a discontinuity surface displaced by sheer forces. Lyell, therefore, happened upon a key element in modern earthquake theory, but he did not develop it systematically.

Most accounts regard the dramatic displacement along the San Andreas Fault after the San Francisco earthquake of 1906 as revealing the key link between faulting and earthquakes. The idea had earlier roots. For example, in 1883 G. K. Gilbert of the U.S. Geological Survey spoke of the earthquake hazard posed by Utah's Wasatch Fault.[7] In Japan, the Nōbi earthquake raised a prominent fault scarp (see fig. 1), which geologist Kotō Bunjirō publicized.[8] Tsuji Yoshinobu points out that one illustration in *Foundation Stone* (*Kaname'ishi*) clearly shows a ruptured fault line in connection with the Kanbun earthquake.[9]

Compared with Japan, the geological sciences advanced further in Europe, especially Britain, by the middle of the nineteenth century. Owing to its high levels of seismic activity and relatively long history of detailed record keeping, Japan could serve as a treasure trove of data. British seismologists, therefore, found Japan a fertile place to work. Italy functioned similarly, and many Western-oriented accounts of the history of seismology locate the beginnings of seismology there. According to Peter M. Shearer, for example, "In 1857 a large earthquake struck near Naples. Robert Mallet, an Irish engineer interested in earthquakes, traveled to Italy to study the destruction caused by the event. His work represented the first significant attempt at observational seismology and described the idea that earthquakes radiate seismic waves away from a focus point (now called a *hypocenter*) and that they can be located by projecting these waves backward to the source. Mallet's analysis was flawed since he assumed that earthquakes are explosive in origin and only generate compressional waves."[10] Knowledge of the history of science in Japan could modify Shearer's assessment. For example, the general idea of seismic waves radiating from a focus point was already known in Japan. Moreover, owing in part to the influence of Dutch publications and in part to the Ansei Edo earthquake, explosive theories of earthquake origins had also become prominent in Japanese academic circles by 1857.

Shearer's account of the history of seismology mentions a few twenti-eth-century Japanese seismologists but says nothing about the origins of seismology in Japan. Nevertheless, it is possible to argue that modern seis-mology was as much an Anglo-Japanese science as anything else. The key development was the arrival in Japan of John Milne (1850–1913) to teach geology and mining in 1876. Milne sought to improve upon Mallet's "obser-vational seismology," which depended on sorting through and measuring wreckage patterns.[11] From the late 1870s, Japan became a major center for seismological research. One advantage Japan had over most other loca-tions was its own seismicity and the vast historical data from the Tokugawa period. The detailed description of damage found in many earthquake accounts could, if well analyzed, serve as a variety of observational seismol-ogy. Initially, of course, Japanese scholars were the only ones in a position to find, compile, and analyze this massive data, a process that continues to this day and has informed this book.

Legacies of Ansei Edo

Interpretations of the Ansei Edo earthquake during the modern era derived from a combination of memorial rites, earthquake lore, earthquakes else-where in Japan felt in Tokyo, and insights provided by seismology. It is useful at the outset to consider the state of seismological knowledge in the last decade of the nineteenth century. The powerful 1891 Nōbi earthquake shook vast areas of central Honshu, and the 1896 Meiji Sanriku earth-quake and tsunami killed approximately twenty-two thousand people in precisely the same area devastated by the March 11, 2011, Great East Japan Earthquake. Although now we usually refer to the 1896 event as an earthquake followed by a tsunami, at the time it was known mainly as a tsunami. Today we know that it was a particularly deadly type of earthquake called a "tsunami earthquake." In such an event, the fault ruptures relatively slowly, producing long-period seismic waves that cause a gentle undulation of the land. Many people experiencing such an event do not realize that it is an earthquake and are therefore less likely to seek high ground in anticipation of a tsunami.[12] With little knowledge of faults and no knowledge of plate tectonics, explaining these events, especially the 1896 tsunami, was a challenge. A brief examination of one of the reports on the tsunami issued by the Imperial Earthquake Investigation Committee illuminates this difficulty.

The fourth chapter of the report is entitled "Discussing the Cause of the Recent Tsunami" (Konkai tsunami no genin o ronzu). It begins with two basic possibilities. Either an earthquake or undersea volcanic activity caused the tsunami. If it was an earthquake, there were two further possibilities. In the thinking of the time, earthquakes whose origin was near the coast in shallow waters usually generate tsunamis, and considerable space is devoted to summarizing past earthquakes of this type, including the Lisbon earthquake, the Ansei Tōkai earthquake, and the Jōgan earthquake of 869. The second type, an earthquake originating in deep water, was mainly a theoretical possibility, so rare that there were no good examples to cite. The report then summarizes the evidence that an earthquake caused the 1896 tsunami and lists four peculiarities of the event that made it atypical of earthquake-generated tsunamis, one of which was "extremely weak ground motion of the earthquake." Therefore, it concludes, the more likely cause was undersea volcanic activity.

Krakatoa's eruption, after all, produced small earthquakes and large tsunami waves that traveled around the world. How is it, though, that large-scale volcanic activity would be located on the sea floor off Japan's Sanriku coast? The report posits, "Indeed there exists a weak point in the earth in this area" because grooves in the earth's crust at this location are aligned in a northeast to southwest orientation. Therefore, volcanic activity could break through the crust in this weakened area and displace enough seawater to cause a tsunami. Moreover, anecdotal reports—for example, by people who were washed into the sea and rescued—indicate that the water temperature was warmer than usual, adding further support to the volcanic activity theory.[13] Not all investigators agreed with the underwater volcanic activity hypothesis. My point in mentioning it here is simply to highlight some of the difficulties of explaining deadly seismic events in the Meiji era. Seismologists of the time could measure ground motion, wave heights, and other metrics with considerable precision. Teams of investigators carefully surveyed and documented the damage. Nevertheless, causal explanations remained highly speculative.

The Ansei Edo Earthquake in Modern Memory

One point to stress at the outset is that the carnivalesque aspects of the Ansei Edo earthquake, so prominent in 1855 and 1856, quickly faded. Even before the end of the Tokugawa period, the earthquake had become solely a terrifying and tragic event. In the back of the minds of many of Tokyo's residents was a fear that an event like Ansei Edo would recur. Smaller earthquakes served to kindle this anxiety. The population looked to the scientific community for reassurance and appropriate warnings in the hope of mitigating future large earthquakes. Anthropologist Susanna M. Hoffman provides a useful framework by characterizing modern catastrophes into two categories: "technological" versus "natural" disasters. Technological disasters are failures of human society, such as the Three Mile Island nuclear meltdown. These disasters "never pass from history to myth," nor do they have any redemptive qualities beyond serving as examples of errors to avoid in the future. Natural disasters, on the other hand, though often horrific at the time of occurrence, are often interpreted as having a good or redemptive side.[14] Hoffman's insights help explain the modern view of Ansei Edo as ominous, despite its earlier interpretation as a redemptive event.

In keeping with its complex, transitional character, the Ansei Edo earthquake was a "natural" disaster at the time of its occurrence and during its immediate aftermath. As we have seen at great length, it was an instance of world renewal to many of Edo's townspeople. Indeed, the major iterations of the earthquake catfish nicely illustrate Hoffman's point that disaster symbolism typically exhibits dualism in the form of creative and benevolent forces (often metaphorically marked as "mother") in tension with destructive and fearsome forces (often marked as "monster").[15] In 1855 or 1856, both aspects were present, but creative themes of renewal were dominant. We have seen, however, that in the intellectual realm the Ansei Edo earthquake prompted a rethinking of the causes of earthquakes and encouraged the view that human agency could predict and mitigate the destructive effects of earthquakes. It is most likely for this reason that early in the Meiji era, Ansei Edo shifted from a natural disaster to a technological disaster. Its destructive power was, at least in part, the result of failure to read the warning signs. It was an error that, hopefully, modern seismologists would help avoid in the future. People perhaps feared that the expectation that seismologists can predict earthquakes was too high a bar, and the Ansei Edo earthquake lurked in the background of modern Tokyo as an ominous possibility, a monster lurking within the earth, not entirely banished or brought under control. One manifestation of this fear was a tendency to exaggerate the 1855 fatality count.

To sketch the legacy of the Ansei Edo earthquake in modern times, I rely mainly on articles appearing in the *Yomiuri shinbun,* a major daily newspaper from the beginning of the Meiji era to the present. One force sustaining public memory of the earthquake was memorial services. The *Yomiuri* announced a Buddhist memorial service in 1877, for example, at Kaikōin in Ryōgoku, to be held the first three days of October. The announcement claimed that the number of earthquake dead amounted to "several tens of thousands" (*nan man*).[16] Other announcements of memorial services appeared from time to time, and some made claims about the nature of the earthquake. One 1887 announcement of services for the thirty-third anniversary of the earthquake stated that approximately ten thousand unidentified victims were buried in a mound at Jōkanji. Because most of these victims were likely to have come from Yoshiwara, the mound is called "Yoshiwara-zuka."[17] In these announcements, the death toll is inflated, albeit moderately. An announcement for the forty-first anniversary services in 1894 stated that "several tens of thousands" (*sūman nin*) died.[18] A 1903 announcement of a forty-ninth anniversary memorial service stated that it would be held in the flower garden of the Yoshiwara brothel district and that the service would feature a procession of children reading sutras.[19] In the context of these memorial services, of course, the sense of world renewal present in the original event was completely absent.

The Ansei Edo earthquake also served as a challenge for the emerging discipline of seismology. It became a baseline against which to measure improvement. The goal for modern science was to provide the knowledge that would lead to better prediction of earthquakes, better recommendations about how to react to earthquakes, and earthquake-resistant construction methods. An announcement for an 1888 meeting of the Earthquake Study Society (Jishin Gakkai), for example, explained that Sekiya Seikei

(1855–1896) would lecture on the Ansei Edo earthquake and that there would be a lecture in English on seismometers and related topics.[20] The Ansei Edo earthquake quickly became an integral part of modern Japanese seismological knowledge.

For this reason, the Ansei Edo earthquake often served as a point of comparison in articles about different aspects of earthquakes. An 1885 article speculating on links between earthquakes and storms featured a dramatic graphic image of the bent spire of Sensōji and implied a link between the earthquake and the typhoon that struck ten months later.[21] Another article that year by Tomisawa Isao on the topic of earthquakes and petroleum speculated that should another earthquake on the scale of Ansei Edo occur, it would destroy all of Tokyo because of widespread petroleum use.[22] With the firestorms of 1923 in mind, Tomisawa's analysis seems prescient.

Most likely owing to aftershocks connected with the October 28, 1891, Nōbi earthquake, some residents of Tokyo reported slight shaking at almost the precise thirty-seventh anniversary of the Ansei Edo earthquake. A November 4 article explained that although the weather was nice, many residents of the capital were on edge, because they thought the air was too warm. The article reported that some nervous residents spent the night in prayer, asking that it pass calmly.[23] The concern with unseasonably warm weather was a legacy of the idea that earthquake shaking was the result of yang energy seeking to escape upward. The warmth came from some of this energy seeping out, and unseasonably warm weather was a commonly cited early modern precursor.[24] Alleged links between weather and earthquakes remains a prominent theme in much of the literature on earthquake prediction.[25]

One result of the powerful Nōbi earthquake was a strong turn toward science, particularly the new science of seismology. In the wake of this earthquake, the Meiji state established its first interdisciplinary scientific body, the Imperial Earthquake Investigation Committee (Shinsai Yobō Chōsakai), the same organization that issued the report on the 1896 tsunami discussed above. One of the hopes of this scientific turn was earthquake prediction, a goal that remains elusive to this day.[26] Articles in the press from this point onward usually took on a scientific tone. Those not authored by seismologists typically referred to the reports and findings of seismologists. Part of the responsibility of the press was to cast off old superstitions and educate the public in the science of earthquakes. In this post-Nōbi context, Ansei Edo remained a common point of comparison.

An 1894 article reported on a "big" earthquake that shook Tokyo, the first real shaking since the Ansei era, though not enough to cause serious damage. The article included data about the epicenter, duration of shaking, direction of shaking, and maximum movement in both vertical and horizontal directions.[27] As the Year of the Rabbit in the old sexagenary cycle loomed in 1915, talk of the Ansei Edo earthquake increased. An article entitled "Earthquake Warning" (Jishin chūi) explained that an old man was overheard on New Year's Day to say that during this year a major earthquake would occur in Tokyo because of what happened sixty years ago. Because a major earthquake had indeed occurred in Italy, the reporter hastened to investigate the matter and discovered the Ansei Edo earthquake. Saying that there may be more in play than simply an old man's superstition, the reporter sought out the esteemed Dr. Ōmori Fusakichi (1868–1923) to ask what people should do in the event of an earthquake. Ostensibly based on Ōmori's advice, because everything tends to collapse in the same direction in an earthquake, one should stay calm, ascertain the direction in which a structure is moving or leaning, and flee in the opposite direction. The safest place to which one could flee is an open area, and standing beside a large tree in an open area offers the highest degree of safety.[28]

A few days after the anniversary of the Ansei Edo earthquake that year, an article claiming statistical backing for the idea of sixty-year earthquake cycles appeared. It went on to explain that major earthquakes are caused by the rupture of faults (dansō), whose point of origin is deep under the ground. It explained that there are two types of ground motion: vertical motion predominates near the epicenter, and horizontal motion predominates at points farther removed. The reason damage in the Nōbi earthquake and an earthquake in Ōmi was especially severe at points removed from the epicenter is because horizontal shaking is more destructive than vertical. The final topic in the article was earthquake-resistant construction, and the main argument was that general quality of construction and the quality of the foundation under a structure are the key factors, not whether a building or house is Western or Japanese in style. The piece ended with a comparison of Japanese and Italian houses, the design of each allegedly being "well matched" to conditions in earthquake countries.[29]

An article in 1917 explaining a presentation by Ōmori conveyed reassurance to Tokyo's residents. Entitled "Tokyo and Earthquakes: No Need to Worry" (Tōkyō to jishin: Shinpai no oyobazu), it sought to calm jittery nerves caused by many recent small earthquakes felt in the capital. Ōmori explained that the origins of these small earthquakes were in the Tsukuba Mountains to the northeast of Tokyo. Moreover, he explained that these earthquakes were entirely different from Ansei Edo, which occurred directly under the city. Therefore, Tokyo's residents need not fear a major earthquake.[30] In view of what would happen less than six years later, such words of reassurance may seem especially ironic or tragic. In the narrow context of the small earthquakes of late 1917, Ōmori's explanation might have been accurate. However, Ōmori frequently encouraged calm with respect to the possibility of a major earthquake in Tokyo, in contrast to his academic rival Imamura Akitsune (1870–1948), who had warned of impending seismic disaster as early as 1905. The 1923 Great Kantō Earthquake marked a shift in seismology away from the meteorology, gathering of statistics, and cartography that Ōmori emphasized toward a greater reliance on mathematics and geophysics. As Clancey points out, the term "Ōmori seismology" became a shorthand for the dark days of a discipline that appeared so dramatically to have failed amidst the ruins of Tokyo in 1923.[31]

Owing to this geophysical turn in Japanese seismology, the Ansei Edo earthquake was no longer as prominent a topic of interest among seismologists after 1923. However, it continued to make appearances in other contexts. For example, soon after the Great Kantō Earthquake a serialized literary drama, *The Great Ansei Earthquake: Revenge of the Daughter* (*Ansei daijishin: Musume no adauchi*), appeared in the *Yomiuri shinbun*. The first installment ran on September 22, and the series continued well into 1924. A November 18, 1923, natural disasters exhibition included the Ansei Edo earthquake. The exhibition mentioned that following that earthquake, grilled catfish had become a popular food item in Edo's restaurants as a symbolic expression of revenge on the forces that caused the earthquake.[32] Indeed, there is some evidence for the eating of grilled catfish at that time.[33]

In Meiji-era newspaper reports, claims of several tens of thousands of deaths in the Ansei Edo earthquake were common. In 1953, the *Yomiuri shinbun* reported on the discovery of a diary written by neighborhood head Utagawa Hoshō. The diary claimed over 110,980 deaths, which the paper reported as fact.[34] The notion that one hundred thousand or more died as a result of Ansei Edo occasionally appears in popular or academic discussion, possibly owing to what we know of the Great Kantō Earthquake, confusion about the various "Ansei" earthquakes owing to loose naming conventions, as well as the exaggerated claims made in 1855.

The January 17, 1995, Awaji-Hanshin earthquake (also known as the Rokkō-Awaji earthquake, the Hyōgo-ken Nanbu earthquake, or in English the Kobe earthquake) killed 6,434 people and came as a surprise to a country whose experts did not expect so powerful (M7.3) an earthquake in that location. In the aftermath of that earthquake, the local and central governments came under criticism for a slow relief response. In academic circles, this criticism had the effect of focusing attention on the Ansei Edo earthquake, which as we have seen was characterized by a fast and relatively effective government relief effort. In other words, the earthquake in Kobe touched off a new wave of academic interest in the Ansei Edo earthquake. One result was the publication of several major works about the social and cultural history of the earthquake that have informed this study. The process whereby earthquakes at one point in time stimulate reflection on those of the past continues.

The Early Modern Legacy in the Modern Era

The 1703 Genroku earthquake was more powerful and more deadly than the Ansei Edo earthquake, but its long-term impact on society was minimal. Indeed, by 1855 writers had to remind Edo's residents that the 1703 event had taken place. By contrast, the social impact of the Ansei Edo earthquake was substantial and long lasting. Ansei Edo figured prominently in Meiji-era debates on architecture and in debates regarding the characteristics of Japan as a nation. Major earthquakes in 1891, 1923, and 1995 stimulated reflection on 1855 in academic and journalistic circles, and its "lessons" partially informed reactions to these later events. For example, Ōmori was in Australia in 1923 when the Great Kantō Earthquake struck. In an interview there, he dismissed reports of as many as one hundred thousand deaths in the Tokyo area. Ōmori's reasoning was that similar figures had circulated in the wake of the Ansei Edo earthquake, despite an actual death toll that was closer to seven thousand.[35]

The Ansei Edo earthquake itself was part of a longer development of earthquake culture that had been under way since the sixteenth century, some aspects of which carried forward. To take a specific example, flashes of light were part of the description of every major earthquake since 1662, and atmospheric phenomena soon became part of discussions of earthquake precursors. The Ansei Edo earthquake reinforced much of this lore and added new items such as claims about magnetic disturbances and catfish, as discussed in chapter 2.[36] Japanese seismologists in the modern era tended to take seriously much of this accumulated earthquake lore and to devote resources to conducting research informed by it.[37] In the paragraphs that follow, I discuss aspects of this earthquake culture, looking backward and forward from the standpoint of 1855.

An article appeared in the *Yomiuri shinbun* in 1915 based on a recent presentation by Imamura. It begins with the popular notion that earthquakes occur in sixty-year cycles, based on the old zodiac cycle. As reported in the article, Imamura confirms that there is some basis to this idea, but he does so by means of a statistical analysis of past earthquakes. Moreover, volcanic eruptions also tend also to follow a sixty-year cycle. Reminiscent of the idea of trapped yang energy beneath the earth, the article explains that major volcanic activity keeps earthquakes mild and vice versa. Moreover, the relative prominence of each type of geological activity alternates, thus producing something more like a 120-year cycle. Because the last major volcanic activity was in the Tenmei era (1781–1789), any earthquakes occurring around 1915 should be mild.[38] According to this article, science has apparently verified the basic truth of lore originally based on cylindrical cycles. From the standpoint of contemporary science, the relationship between volcanic activity and earthquakes, if any, is unclear.[39] The tendency to assume statistical regularity for seismic events continues to this day in what are referred to as "characteristic earthquakes."

A particularly significant figure with respect to the modern influence of Tokugawa-period knowledge and lore is Musha Kinkichi (1891–1962). Musha was a seismologist who invested considerable time and energy into compiling historical documents connected with earthquakes, following up on similar efforts by Sekiya Seikei and Ōmori. Musha was a representative of the "Ōmori seismology" that lost favor after 1923. Until the end of his life, Musha argued that researchers should take seriously and further investigate certain phenomena well attested in early modern documents. He was also convinced that with more effort, earthquakes could be both predicted and prevented. For example, he thought it difficult but possible to figure out a way to release energy (*enerugii*) that had built up under the earth's crust and posed an earthquake danger.[40]

In chapter 2 we saw that early modern earthquake signs included unseasonably hot weather; thunder; well water increasing, decreasing, or becoming muddy; stars appearing closer, brighter, or reddish in color; flashes of light; and, after 1855, strange behavior of catfish. Musha was not persuaded that all of these phenomena were worthy of attention. He seemed convinced,

however, that when there was overwhelming documentary record of a phenomenon, to ignore it was tantamount to accusing our ancestors of making
up lies. Today psychologists are well aware of the phenomenon whereby
people think they see what they expect to see. As Thomas Gilovich explains,
"When examining evidence relevant to a given belief, people are inclined
to see what they expect to see, and conclude what they expect to conclude.
Information that is consistent with our pre-existing beliefs is often accepted
at face value, whereas evidence that contradicts them is critically scruti-
nized and discounted."[41] Closely related to this tendency for people "to see
in a body of evidence what they expect to see" is what psychologist Kikuchi
Satoru has called "mistaken correlation" (sakugo sōkan). For some people
with preconceived notions about the nature of certain events, the occurrence of that event suddenly makes ordinary, mundane phenomena seem
significant in hindsight. In extreme cases, people can manufacture earthquake "precursors" in the wake of happenings vaguely remembered.[42] In
his writings, Musha seems entirely unaware of the possibility that people
expected to see certain things in earthquakes and therefore thought that
indeed they had seen these things. To take one example that remains influential to this day, Musha was convinced that fish behavior could help predict earthquakes. Here I examine this matter in some detail as a case study
of the process whereby early modern lore has influenced modern and contemporary scientific research and popular thought about earthquakes.

The basic idea was that certain fish might be unusually sensitive to
some change that occurs just before an earthquake. The most common
line of reasoning assumes that there must be some kind of electromagnetic
change just prior to an earthquake. This point is unproven but commonly
supposed, often with the alleged magnetic disturbances preceding Ansei
Edo cited as evidence. Another possibility was that catfish are sensitive to
vibrations from seismic waves too faint to be measured by instruments,
which might precede the main shock of an earthquake. Research conducted
during the 1920s and 1930s and between 1977 and 1993 has suggested that
indeed, catfish are especially sensitive to electrical currents and vibrations.[43]
An April 1, 1932, Yomiuri shinbun article announced with fanfare that
Tōhoku University professor Hatai Shinkishi had demonstrated that when
catfish swim in a certain way, an earthquake occurs within twelve hours
and that the fish in his lab had predicted almost a hundred earthquakes.[44]
One problem, however, is that small earthquakes occur very frequently,
so it was unclear whether the catfish really were sensing earthquakes or

whether their swimming and the earthquakes were coincidental. Nearly forty-five years later, a February 1977 headline in the same newspaper read, "Leave Earthquakes to Catfish" (Jishin nara namazu ni makaseru). The article explained a government-funded project to place both American and Japanese catfish in aquariums to serve as "an advance guard of earthquake prediction" constantly on duty. The basic setup was a device in the tank that would detect movement and sound an alarm. At a total cost of 120 million yen (US$1.5 million at the July 2012 exchange rate), the program continued until March 1993. Its advocates claimed a 30 percent "success" rate, though not in a strict sense of specifying in advance when, where, and how strong an impending earthquake would be. Instead, the standard was that the fish exhibited odd behavior within ten hours of an M3 or higher earthquake.[45] So far, nobody has created a practical scheme whereby fish might usefully predict earthquakes. In summarizing the scientific data on catfish, Rikitake Tsuneji wrote circa 1995 that the idea of fish as earthquake predictors is "not absurd" (baka ni shita mono de wa naku) and should be considered in future research on earthquake prediction.[46] At the time the catfish research was shut down in 1993, Rikitake said, "In ancient times and the present, east and west, people have reported such things as dogs barking before earthquakes. Even if only one in a hundred such reports are accurate, we should pursue them."[47]

Musha enthusiastically praised the work of Hatai and the ichthyologist Suehiro Yasuo of the Imperial Fisheries Experimental Station. As for Hatai's work with catfish, Musha claimed the correlation between the fish becoming agitated and the occurrence of an earthquake within fifteen hours or less (eight hours was typical in the case of the strongest earthquakes) was so strong that it could not have been attributable to chance. The most likely mechanism linking the catfish and earthquakes was a subtle change in voltage, and catfish can detect a change in electrical current as small as one microampere. Therefore, Musha concluded, the folkloric connection between earthquakes and catfish was not baseless superstition.[48] Despite his familiarity with historical records, Musha did not mention that the specific idea that catfish could usefully predict earthquakes came from a tale of unknown provenance in 1855 and does not seem to have been prevalent earlier.

Expanding his discussion, Musha pointed out that both Sekiya and Ōmori conducted research on pheasants and earthquakes (pheasants sometimes substituted for catfish in early modern lore) but that fish have not received adequate academic attention. John Milne published an essay on earthquakes and lower animals in 1888, but it took over forty years for Imamura to write about the subject, even then barely mentioning fish. The publication of Hatai's findings, Suehiro Yasuo's research in the wake of the 1933 Shōwa Sanriku earthquake, and an article on the strange behavior of fish in the wake of the Oga earthquake of 1939 finally brought fish into the academic spotlight. As a whole, this research reveals four types of behavior that fish might exhibit prior to an earthquake: (1) strange behavior; (2) gathering in a group at the water's surface and frequently jumping; (3) gathering near rocks; and (4) certain species temporarily disappearing.[49]

The following sample includes only a few of Musha's many examples. Just before the Nōbi earthquake, large numbers of loaches appeared in the wet fields of Gakuden Village in Aichi Prefecture. Before the Sanriku tsunami of 1896, vast numbers of eels gathered along the coast, and even in the daytime, normally secretive eels protruded from their holes. Prior to the Great Kantō Earthquake there were reports from the waters off the Izu coast that fish called *shige* (probably *suketōdara* [Alaska pollock], *Theragra chalcogramma*), which normally inhabit deep waters, could be found floating at the surface. Moreover, the Shōwa Sanriku earthquake occurred around 2:30 a.m., but Suehiro discovered that four and a half hours later at a beach near Odawara, fishermen caught a rare ribbonlike threadfish (*Nemichthys avocetta*) that normally lives only in waters three thousand meters deep. Suehiro concluded that the fish must have sensed some change in the focal area of the earthquake and fled at great speed, reaching the waters of Kanagawa Prefecture by around 7:00 that morning. Moreover, the stomach contents of sardines caught the day before the earthquake contained plankton normally found only at the bottom of the sea. Therefore, even plankton rose upward, perhaps in an attempt to get away from the earthquake.[50] Suehiro, later known as "Dr. Fish" (O-Sakana Hakase), was cited in the 1977 *Yomiuri* article on catfish claiming that unusual fish behavior was linked with earthquakes.[51]

Musha and others claim that prior to the 1896 tsunami many sardines gathered along the Sanriku coast, a phenomenon also reported prior to a tsunami in 1856. Moreover, in 1856, 1896, and 1933, fishermen along the Sanriku coast pulled in vast quantities of sardines just before the earthquakes and tsunamis struck, and in the aftermath of the tsunamis, they caught unusually large quantities of squid. Conversely, in 1896 just prior to the tsunami, fishermen could catch no cod or sharks, and in 1933, sea cucumbers became almost impossible to find just before the earthquake and tsunami. Moreover, prior to the Great Kantō Earthquake, fish stopped biting in Sagami Bay.[52]

Speculating on the reasons for this wide range of strange fish behavior, Musha proposed that factors beyond changes in electric currents must be at play. He quoted Imamura theorizing that fish can detect precursor earthquakes too small to be detectable even by instruments. Upwellings of subterranean water or even oil might also play some role. Musha reported that seismologist Terada Torahiko (1878–1935) proposed three possibilities: (1) earthquake motion produces machinelike stimulation that fish can sense; (2) earthquake movement has an effect on organisms like plankton that serve as food for fish and thus indirectly affects fish behavior; and (3) earthquakes cause changes in water chemistry near the coast that affect the behavior of fish and plankton.[53]

Following this discussion, Musha says of fish and earthquakes, "Without experimental proof, however, it is impossible to convince everyone. It is regrettable that as of now, we cannot go beyond the bounds of speculation."[54] Presumably, the correlation that Musha found in Hatai's experiments, allegedly too high to be attributable to chance, was not sufficient to convince many other scientists. In light of Musha's statement, it is worth considering a point from the psychological literature: "We humans seem to be extremely good at generating ideas, theories, and explanations that have the ring of plausibility. We may be relatively deficient, however, in evaluating and testing our ideas once they are formed. One of the biggest impediments to doing so is our failure to realize that when we do not precisely specify the kind of evidence that will count as support for our position, we can end up 'detecting' too much evidence for our preconceptions."[55] Suehiro's threadfish, for example, could have come from anywhere. The idea that it managed to swim all the way from off the Sanriku coast to a beach south of Tokyo in a few hours could only seem to be plausible evidence to someone already convinced that fish can anticipate earthquakes and that they behave much like humans would—that is, they try to escape the zone of shaking. Similarly, consider the logic in theorizing Hatai's catfish. How did anyone know that a change in electrical current precedes earthquakes? This point has never been established in a rigorous scientific manner. The idea fit with older lore about magnets and electricity, but if instruments of the time could have measured it, there would have been no need for the fish. Musha and others display a circular reasoning. They started with the axiom that fish can detect earthquakes. If so, there must be a reason, and we know that catfish are sensitive to electricity because they use a sonarlike system to locate prey. The speculation about electrical changes in the earth or subtle precursor seismic waves served to explain what we already know about catfish, namely their ability to sense earthquakes. This process has generated so much discussion of fish and earthquakes that its sheer quantity is sufficient for many people to think that there must be something to the idea.

Musha's discussion goes on to include changes in well water, which might also explain fish behavior, and the sensitivities of rats, cats, dogs, monkeys, horses, donkeys, various birds, snakes, and frogs to seismic activities.[56] In the course of summarizing the vast quantity of reports regarding the behavior of fish and other animals, Musha returns to the same basic point he made in connection with other alleged precursors not taken sufficiently seriously by the scientific community: "It is hardly the case that both Japanese and foreigners consulted with each other and wrote down lies. It is hardly reasonable to regard every one of the extensive documents presented here as falsifications."[57] It is useful here to consider the issue of large quantities of poor-quality evidence. In the context of discussing the widespread popular belief in extrasensory perception (ESP), a belief similar in many ways to earthquake precursors, Gilovich explains, "Much of the evidence may be fraudulent or faulty, but there is so much of it. Can't we conclude that 'where there's smoke, there must be a fire?' We cannot, of course. . . . But this reasoning is seductive nonetheless. Indeed, it is not just the average person who falls prey to this fallacy. Well-trained scientists have been known to make the same argument."[58] The fish stories and other tales that came out of the Ansei Edo earthquake made "sense" in their original context. Amazing earthquake tales helped sell books like *Ansei Record* and *Ansei Chronicle*, originally commercial publications for mass audiences that have now acquired the status of venerable historical records. Moreover, early modern lore about weather patterns and other precursors derived from a conception of earthquakes that no scientists and very few nonscientists held by the turn of the twentieth century.

The topic of animal behavior and earthquakes seems to come up in the popular media in the wake of every major earthquake, but the experimental proof for which Musha longed has not been forthcoming. Tales of aquatic creatures and earthquakes continue to circulate in the mass media, even if most scientists have lost interest. We have seen in the introduction that the Great East Japan Earthquake of 2011 has generated a new round of aquatic tales and even prompted the City of Susaki seriously to consider relying on animal behavior as a defense against earthquakes. No doubt, this phenomenon of revisiting old lore is in part the result of frustration at the inability of seismologists to predict earthquakes, particularly in Japan, where law and government policy explicitly state that scientists must endeavor to predict earthquakes, supported by taxpayer money.[59] Both the 1995 earthquake in Kobe and the 2011 disaster occurred without the slightest prior warning, despite what is probably the most concentrated array of earthquake detection and measuring devices in the world. Public expectations of science are often unrealistic, and just like a patient diagnosed with an incurable disease, there is an understandable tendency to seek alternatives when science cannot provide clear answers. In the case of earthquakes in Japan, the large accumulation of lore from the early modern era provides a rich storehouse of what we might call "alternative seismology."

Japan as a Modern Earthquake Country

Clancey explains that seismology in Japan "was not just subterranean science." It was closely intertwined with questions of national character and national essence. Japan's seismicity had direct implications for architecture, and particular conceptions of Japan as an "earthquake country" competed to determine the physical appearance of urban spaces. "Seismology and architecture had this in common: both sought to explain national character with reference to geology."[60] The Meiji period was also a time when many Japanese began to ponder the broader question of how their country's natural features and landscape shaped what was assumed to be its unique culture. Especially instrumental in what we might call a "geographical turn" was Shiga Shigetaka (1836–1927), whose 1894 book, *On Japan's Landscape* (*Nihon fūkeiron*), became a best seller. Shiga emphasized volcanoes in shaping Japan's national character, but the approach he popularized also lent urgency to the work of seismologists. Writing in the periodical *Nihonjin* (*The Japanese*), which Shiga edited, he stated, "The influence of all environmental factors of Japan—her climate and her weather conditions, her temperature and humidity, the nature of her soil, the configuration of her land and water, her animal and plant life and her landscape, as well as the interaction of all these factors, the habits and customs, the experiences, the history and development of thousands of years—the totality of all these factors has gradually, imperceptibly, developed in the Japanese race inhabiting this environment a unique *kokusui* [national essence]."[61] Shiga engaged European writers such as Thomas Buckle, rejecting his idea that gentler landscapes meant greater progress. Shiga was familiar with the work of John Milne and argued that volcanic Italy, cradle of Western civilization, was an important point of comparison for Japan.[62] Indeed, the early modern tendency to compare Japan and China with respect to earthquakes shifted in the Meiji era to constant comparisons of the two "earthquake countries," Japan and Italy.[63]

The Great Kantō Earthquake occurred precisely at a high point in social anxieties about modernity, urbanization, changing gender roles, and related matters. Therefore, it produced extensive discussion of divine retribution, possibly even to a greater extent than most early modern earthquakes. Of course, it is possible that many such characterizations of that earthquake deployed divine retribution as a figure of speech, not as a reflection of theological belief. Despite continuities with the early modern past in rhetoric and in earthquake-related lore, the Meiji era marked a major shift in conceptions of Japan and the significance of earthquakes. Early modern earthquakes caused nervous survivors to remind themselves that Japan is a resilient *shinkoku,* blessed by thousands of benevolent albeit somewhat underpowered deities. In the Meiji era, earthquakes reminded Japanese that they lived in a country subject to frequent violent outbursts of nature. Furthermore, there was a strong tendency to assume that this severe natural environment has molded Japan's people, architecture, culture, and national character, even if the precise qualities were debatable and elusive.

Somewhat reminiscent of the Tokugawa era was the capacity of earthquakes to serve as moral drama. As we have seen, works such as *Ansei Record* and *Ansei Chronicle* claimed an edifying purpose, even if their entertainment value is what sold copies. Modern earthquakes and other disasters produced precisely the same kind of moral drama, but in a more systematic manner and often with support from the state. Janet Borland gives the following example that took place after the Great Kantō Earthquake :

During September and October, Boy Scouts wandered throughout Tokyo and Yokohama in search of appropriate stories. They interviewed people ranging both in age and social status, from young schoolboys and girls, to servants, policemen and school teachers, so as to record miracles of survival and escape, as well as tragic tales of sacrifice, suffering and loss of life. In particular, they sought stories which characterised praiseworthy qualities of the ideal Japanese subject. Although the assignment was made difficult by the grim conditions prevailing in Tokyo, the Boy Scouts completed the task with great efficiency so that within a matter of weeks the Ministry of Education began to sort, select, and edit the collection of stories for distribution.[64]

Such a project bore some resemblance to *Ansei Record* or *Ansei Chronicle*, but it was different in significant respects. The primary purpose in 1923 was edification, not entertainment. Moreover, in 1855 a wide range of behavior, good and bad, and a range of outcomes, just and unjust, emerged in the pages of Ansei Edo earthquake accounts. By contrast, *Materials for Education in Connection with the Earthquake* (*Shinsai ni kansuru kyōiku shiryō*), the Ministry of Education's earthquake-related educational materials, focused exclusively on exemplary stories. Only the wide range of subjects covered and the inclusion of an account of the damage was reminiscent of the 1855 earthquake accounts.[65] The Great Kantō Earthquake functioned to highlight the exemplary character of the Japanese people.

Similar phenomena took place in the wake of other twentieth-century catastrophes, albeit without direct state involvement. For example, Timothy Tsu explains what occurred after a devastating 1938 flood in Kobe:

> A metropolis renowned for its cosmopolitan culture, modern port, thriving economy, and strategic importance lay in ruin. Against this grim backdrop, and as soon as the floodwater had receded, the local daily *Kōbe shinbun* began to carry "beautiful tales" [*bidan*] of bravery that had saved lives and kindness that had sustained needy survivors. Official "flood records" [*suigaishi*] published in the following years offered more such heart-warming episodes. These purportedly real-life dramas celebrate instinctive expressions of human fellowship. The courage, generosity and usually fortuitous outcomes they narrate stand in stark contrast to the reality of death and destruction in the wake of the deluge.[66]

Tsu further argues that "the beautiful tales from Kobe, taken together, constitute a larger narrative that aims to transform an otherwise horrific event into a self-affirming experience for the city and the nation."[67] Here, too, the coordination of a central editing body helped narrow and focus the range of messages.

Similarly, tales of filial devotion and other ideal virtues emerged from the Shōwa Sanriku earthquake. "The father and daughter who sacrificed themselves to the filial path," for example, is the story of Oda Kiyohachi, head of Tanohata Village in Iwate Prefecture. As a massive tsunami moved toward his residence, Oda calmed his family, put his aged mother on his back, and had his daughter lead her aged grandfather by the hand. Later, the bodies of all four were found, the mother still on Oda's back and his daughter still grasping the hand of her grandfather. The sight moved survivors to tears.[68] Readers of such tales could vicariously partake of both their emotional impact and demonstrations of the "loyalty and filial piety" that by the 1930s had become rhetorical hallmarks of the moral fiber of true Japanese. Relative lack of state control and a central guiding ideology made the moral stage of 1855 Edo a more varied and confusing place than the moral stages of 1923 Tokyo, the 1933 Sanriku coast, or 1938 Kobe.

Although there is much more to say about many of the topics raised in this chapter, it should be clear that early modern earthquake culture and lore, and the Ansei Edo earthquake in particular, have continued to affect the modern and contemporary worlds in realms as diverse as seismology, politics, popular culture, and national images. Located at or near the intersection of four tectonic plates, Japan is among the world's most seismically active countries. Although this seismicity may not have created Mt. Fuji and Lake Biwa in the manner suggested by early modern texts, it has contributed to the making of Japan in many other ways. The process continues.

Postscript: Rhetoric after the Great East Japan Earthquake

Any systematic discussion of an event as large as the March 11, 2011, Great East Japan Earthquake, tsunami, and nuclear disaster is well beyond the scope of this study. Indeed, sorting through the data and assessing the larger significance of this event will be an ongoing project for years to come. From this study, it should be clear that the Sanriku coast of northeastern Japan has been no stranger to large earthquakes and deadly seismic sea waves. Indeed, the 1896 tsunami still holds the dubious distinction of being the most deadly tsunami in Japan's history. The subduction zone off the Sanriku coast was a well-known "earthquake nest," at least among seismologists. Had an earthquake of M7 or M8 occurred at the same place and the same time, it would still have been deadly and destructive, but the event would have been within the pale of expert imagination. The occurrence of an M9-class earthquake took many, but not all, experts by surprise. Moreover, because much of the seismological community, politicians, and the public have been fixated for decades on the possibility of a repeat of the 1854 Ansei Tōkai/Nankai earthquakes and tsunamis, the 3/11 disaster was even more of a surprise. The nuclear disaster accompanying this event added a new dimension of apprehension to the massive death and destruction caused by the wave trains, which reached nearly forty meters in some places. Indeed, for these reasons, it may even be reasonable to call this multifaceted disaster "unprecedented," at least in terms of its effects.

The element of surprise became the major defense of TEPCO officials and politicians. It was "an act of God" (*kamisama no shiwaza*), said acting finance minister Yosano Kaoru in May 2011. Others offered similar explanations for their lack of preparedness, characterizing the event as "beyond anything we could have supposed" (*sōteigai*). In a scathing critique of such excuses, seismologist Robert Geller characterizes the nuclear disaster as an act not of God but of "negligence" or "omission." Megathrust earthquakes in Kamchatka (1952, M9.0), Alaska (1964, M9.2), Chile (1960, M9.5), and Sumatra (2004, M9.3) should have prompted some consideration of the possibility of such an event occurring in one of Japan's subduction zones. Furthermore, an M9-class earthquake in the Cascadia Subduction Zone probably generated the "orphan tsunami" of 1700, and the Jōgan earthquake of 869 that struck Japan's Sanriku coast was probably an M9-class event, described in the *Journal of Geology* in 1991 and other publications in subsequent years. Moreover, several experts in the years before the 3/11 disaster specifically warned of an M9-class earthquake and tsunami causing disastrous consequences for nuclear plants. Policy makers and power company officials ignored them.[69]

The public discourse on 3/11 is still evolving, although some aspects of it are clear. Here I briefly examine rhetoric in the wake of 3/11 in light of the early modern post-earthquake rhetoric we have examined. Not only did every major Tokugawa-era earthquake generate private and public materials that characterized the event as "unprecedented," this term also appeared frequently in connection with modern earthquakes and tsunamis in 1891, 1896, 1923, and 1933. In some respects, the rush by those potentially responsible for the human aspects of the disaster to declare the event outside the bounds of the imagination resembles the earlier declarations of earthquakes as unprecedented, albeit with much more self-interest at stake. Rebukes pointing out that TEPCO and nuclear regulatory authorities should indeed have been able to imagine an M9 earthquake inevitably invoke the seismic past of Japan and surrounding areas.

Characterizing 3/11 as unprecedented would not work well in our information-rich age. Anyone willing to do a modicum of work with a search engine would come to know about the large tsunamis that washed over the Sanriku coast in 1611, 1857, 1896, and 1933. Moreover, by the time of 3/11, the Jōgan earthquake had become widely discussed as a likely M9-class event, and the other M9 earthquakes around the Pacific had been known since the advent of moment magnitude in the late 1970s. If characterizing 3/11 as unprecedented would be implausible, perhaps it could still be regarded as beyond the reasonable limits of the imagination. The discipline of paleoseismology provided the needed term.

Within months after 3/11, paleoseismologists announced the discovery of evidence of an M9-class event in northeast Japan during the Yayoi period (roughly 300 BCE–300 CE) and then evidence of a series of such events averaging out to roughly one-thousand-year intervals of occurrence.[70] This idea of a thousand-year interval, which originated in paleoseismology, soon become part of popular discourse. In the process, it lost a substantial part of its original meaning. The current term is "1,000-year earthquake" (*sennen jishin*), which functions much like "unprecedented" did in earlier earthquake discourse. Just as "unprecedented" was often a figure of speech to emphasize the sense of enormity of the seismic event in question, so too has the term "1,000-year earthquake" quickly taken on flexible meanings. For example, the main title of a recent book discussing a variety of early modern and modern earthquakes is called *Sennen shinsai* (1,000-year shaking disaster), a slight variant on this recently popular term.[71] In other words, any large earthquake in the past might now be called a "1,000-year earthquake," at least in nontechnical contexts.

Insofar as M9-class earthquakes affecting northeast Japan may indeed average roughly one thousand years, emphasizing this point surely shifts focus away from the human contribution to the catastrophe. After all, who could have realistically anticipated an event occurring so rarely? Such a focus helps obscure the fact that an M8-class event on the scale of 1896 or 1933 would have caused vast damage and loss of life, albeit not to the full extent of what occurred in 2011. Moreover, as we have seen, if we take the Pacific region as a whole as our geographical focus, several M9-class earthquakes have occurred in recent decades, some close to Japan. One final

point is that even if a thousand years does prove to be the average interval of M9-class earthquakes in northeastern Japan, it does not necessarily
mean that another such event will wait for another thousand years. Sudden,
violent movements of the earth are notoriously unpredictable, as we have
seen throughout this study.

NOTES

Chapter 1 Earthquakes in the Early Modern Era

1. Daiki Yamamoto, "Sea Serpents' Arrival Puzzling, or Portentous?" *Japan Times,* March 6, 2010.
2. "Ika no toresugi wa daijishin no zenchō? Tokushima de 4-bai mo," *Yomiuri Online,* May 1, 2011.
3. Kashima Tarō [pseudonym], *Kashima no kami to Suwa no kami: Higashi Nihon dai shinsai no fukkō ni mukete* (Kashima-shi, Japan: Kashima jingū shamusho), 2011.
4. "Shinkaigyo, Shizuoka-ken de aitsugi shutsugen . . . daijishin no maebure?" in *Yomiuri shinbun,* June 9, 2012.
5. Yoshimura Akira, *Sanriku kaigan ōtsunami* (Bungei shunjū, 2004), 16–20, 81–82.
6. "City Looks to Base Tsunami Warnings on Animal Behavior," *Japan Times,* June 3, 2012; "Japanese City to Watch Animal Behaviour for Disaster Signs," *Tokyo Times,* 2012.
7. For a critique of the idea of characteristic earthquakes (*koyū jishin*), see Robert Geller (Robaato Geraa), *Nihonjin wa shiranai "Jishin yochi" no shōtai* (Fatabasha, 2011), 159–165, and Yan. Y. Kagan, David D. Jackson, and Robert J. Geller, "Characteristic Earthquake Model, 1884–2011, R.I.P.," in *Seismological Research Letters* 83 (November/December 2012), 951–953.
8. Shinsai yobō chōsakai, eds., *Dai-Nihon jishin shiryō* (hereafter *DNJS*), vol. 2 (*otsu*), esp. 564–585, and Suzuki Tōzō and Koike Shōtarō, eds., *Fujiokaya nikki* (hereafter *FN*), esp. 513. These versions of *Honchō jishin no shidai* are slightly different, but both claim seismic activity as the creative force for Lake Biwa and Mt. Fuji.
9. "Kainai jishinroku," in Tōkyō daigaku jishin kenkyūjo, ed., *Shinshū Nihon jishin shiryō* (hereafter *NJS*), vol. 5, supplement 2, part 1, 503–506.
10. 120 million yen would be about 1.5 million U.S. dollars in July 2012. The catfish project produced no useful results. For more details, see Geller, *Jishin yochi,* 120–135.
11. For a typical example, see Motoji Ikeya, *Earthquakes and Animals: From Folk Legends to Science* (River Edge, NJ: World Scientific, 2004). Ikeya's main argument is that electromagnetic fields are earthquake precursors and that a variety of animals can detect them. One caption in the front matter reads, "Nails about to drop from a magnet on introduction of an electric field,

reproducing an event known to have occurred two hours before the Ansei Edo Earthquake in 1855" (vi). As we will see, however, there is no evidence that such an event actually took place.

12. Regarding the Oakland firestorm of 1991, which wiped out thousands of affluent homes, some flatlanders expressed the view that excessive consumption invited divine retribution. "Those people in the hills deserved to be wiped out. It was God acting," said one interviewee. Susanna M. Hoffman, "The Monster and the Mother: The Symbolism of Disaster," in Susanna M. Hoffman and Anthony Oliver-Smith, eds., *Catastrophe and Culture: The Anthropology of Disaster* (Santa Fe, NM: School of American Research Press, 2002), 132.

13. Chen Fu, *Nongshu* (Taibei: Taiwan shangwu yinshuguan, 1956), vol. 1, 10–12.

14. Anna A. Akasoy, "Interpreting Earthquakes in Medieval Islamic Texts," in Christof Mauch and Christian Pfister, eds., *Natural Disasters, Cultural Responses: Case Studies toward a Global Environmental History* (New York: Lexington Books, 2009), 190–192.

15. Akasoy, "Medieval Islamic Texts," 189–190.

16. Gregory Quenet, "Earthquakes in Early Modern France: From the Old Regime to the Birth of New Risk," in Andrea Janku, Gerrit Schenk, and Franz Mauelshagen, eds., *Historical Disasters in Context: Science, Religion, and Politics* (New York: Routledge, 2012), 100–103.

17. Elaine Fulton, "Acts of God: The Confessionalization of Disaster in Reformation Europe," in Andrea Janku, Gerrit Schenk, and Franz Mauelshagen, eds., *Historical Disasters in Context: Science, Religion, and Politics* (New York: Routledge, 2012), 62, 66.

18. Quenet, "Early Modern France," 103–105.

19. Ibid., 98.

20. Ibid., 109.

21. Quenet, "Early Modern France," 107–109, and Benjamin Reilly, *Disaster and Human History: Case Studies in Nature, Society, and Catastrophe* (Jefferson, NC: McFarland & Company, 2009), 78–79.

22. Quoted in Quenet, "Early Modern France," 109–110.

23. Christof Mauch, "Introduction," in Christof Mauch and Christian Pfister, eds., *Natural Disasters, Cultural Responses: Case Studies toward a Global Environmental History* (New York: Lexington Books, 2009), 4.

24. Mauch, "Introduction," 9.

25. Anthony Oliver-Smith, "Theorizing Disasters: Nature, Power, and Culture," in Susanna Hoffman and Anthony Oliver-Smith, eds., *Catastrophe and Culture: The Anthropology of Disaster* (Santa Fe: School of American Research Press, 2002), 27–28.

26. Ibid., 30.

27. Perhaps the most influential such work was Wadatsumi Kiyoshi, *Zenchō shōgen 1519: Hanshin-Awaji daishinsai 1995 nen 1 gatsu 17 nichi gozen 5 ji 46 fun* (Tōkyō shuppan, 1996).

28. For a critical assessment of modern fears of a Tōkai or Nankai earthquake, see Geller, *Jishin yochi,* 10–11, 92–98. According to Geller, mentioning the four characters "Tōkai earthquake" functions as "a magic mallet" (*uchide no kozuchi*) that produces budget allocations. The Great East Japan Earthquake and tsunami renewed fears of Tōkai and Nankai earthquakes. To take one example from popular media, see "Massive Tsunami Projected: Panel Forecasts Nankai Trough Quakes Could Affect 11 Prefectures," *Daily Yomiuri Online,* April 2, 2012.

29. Okada Yoshimitsu, *Saishin Nihon no jishin chizu* (Tōkyō shoseki, 2006), 139–143, and Sangawa Akira, *Jishin no Nihonshi: Daichi wa nani o kataru ka* (Chūōkōron shinsha, 2007), 34–36. Nankai Trough earthquakes took place in 684, 887, 1009, 1361, 1605, 1854, and 1946. These earthquakes have often occurred in close temporal proximity to ocean trench earthquakes originating in the Sagami Trough off the Izu Peninsula, which are known as Tōkai earthquakes. They occurred in 1096, 1498, 1605, 1707, 1854, and 1944. For more details on Tōkai and Nankai earthquakes, see Tsuji Yoshinobu, "Ansei tōkai, nankai jishin no jitsuzō to senjin no saigai kyōkun," in Chūō bōsai kaigi, *1854 Ansei tōkai jishin, Ansei nankai jishin hōkokusho* (Chūō bōsai kaigi, 2005), 4–5, and Noguchi Takehiko, *Ansai Edo jishin: Saigai to seiji kenryoku* (Chikuma shobō, 1997), 22.

30. Itō Kazuaki, *Jishin to funka no Nihonshi* (Iwanami shoten, 2002), 184; Okada, *Jishin chizu,* 144; and Usami Tatsuo, *Nihon higai jishin sōran [416]–2001, saishin-ban* (Tōkyō daigaku shuppankai, 2003), 52.

31. Okada, *Jishin chizu,* 144, Itō; *Jishin to funka no Nihonshi,* 182, 184; and Usami, *Higai jishin,* 51–52. For reports on damage to the castle, see *DNJS,* vol. 1 (kō), 157–158, 191. For the specific figures of seventy-three ladies-in-waiting and five hundred maids killed, see pages 202–203.

32. Okada, *Jishin chizu,* 145; Itō, *Jishin to funka no Nihonshi,* 185; Sangawa, *Jishin no Nihonshi,* 121–123, 125; Usami, *Higai jishin,* 57–58; and Kitahara Itoko, "Saigai to jōhō," in Kitahara Itoko, ed., *Nihon saigaishi* (Yoshikawa kōbunkan, 2006), 232. For a detailed analysis of the damage from this earthquake, see Chūō bōsai kaigi, *1662 Kanbun Ōmi-Wakasa jishin hōkokusho* (Mizuho jōhō sōken kabushikigaisha, 2005). For extensive diary and other accounts of the earthquake and its damage, see *DNJS,* vol. 1 (kō), 246–255. Regarding the buried villages, see page 251 (entry "Rakuho zatsudan ichigenshū").

33. Okada, *Jishin chizu,* 86–87; Itō, *Jishin to funka no Nihonshi,* 80–91; Sangawa, *Jishin no Nihonshi,* 134–137; and Usami, *Higai jishin,* 65–70. For diaries,

official chronicles, and other accounts of the earthquake and its damage, see *DNJS*, vol. 1 (kō), 281–307.

34. Sangawa, *Jishin no Nihonshi,* 136. For a detailed account of damage to daimyō and *hatamoto* mansions, see "Kanrosō," in *DNJS*, vol. 1 (kō), 290.

35. Okada, *Jishin chizu,* 169–170; Sangawa, *Jishin no Nihonshi,* 138–141; Itō, *Jishin to funka no Nihonshi,* 81; and Usami, *Higai jishin,* 75–90.

36. Okada, *Jishin chizu,* 56; Sangawa, *Jishin no Nihonshi,* 158–162; Usami, *Higai jishin,* 130–131; and Hashomoto Manpei, *Jishingaku no kotohajime: kaituakusha Sekiya Seikei no shōgai* (Asahi shinbunsha, 1983), 40.

37. "Chōshin hiroku, jōkan," in Usami Tatsuo, ed., "Nihon no rekishi jishin shiryō" shūi (hereafter *NRJSS*), vol. 3, 211.

38. Sangawa, *Jishin no Nihonshi,* 162; Itō, *Jishin to funka no Nihonshi,* 187–190; Usami, *Higai jishin,* 131–132; and Miki Haruo, *Kyōto daijishin* (Shibunkaku shuppan, 1979), 4–48, 115–250. Miki argues that an accurate count of the dead is impossible. See also *DNJS*, vol. 1 (kō), 533–589.

39. Musha kinkichi, *Jishin namazu* (Meiseki shoten, 1995, originally 1957), 55.

40. For an extended discussion of the causes and other seismological details, see Tsukahara Hiroaki, "Zenkōji jishin wa dono yō nishite hasseishita ka," in Akahane Sadayuki and Kitahara Itoko, eds., *Zenkōji jishin ni manabu* (Nagano-shi, Japan: Shinano mainichi shinbunsha, 2003), 35–47.

41. Okada, *Jishin chizu,* 118–119; Itō, *Jishin to funka no Nihonshi,* 128–142; Sangawa, *Jishin no Nihonshi,* 164–170; Kitahara, "Saigai to jōhō," 246–246; Usami, *Higai jishin,* 137–144; and Shinano mainichi shinbunsha kaihstsukyoku shuppanbu, ed., *Kōka yonen Zenkōji daijishin* (Nagano-shi: Shinano mainichi shinbunsha, 1977), 111–204. For documentation of the changing condition of the Sai River and other rivers, see "Shinetsu jishinki," 9, 11–12; "Fujikawa Kan zakki," 46–47; "Kamahara Dōzan jishin kiji," 71–78, 92–93; "Kenshūroku," 105–107, 109–111, 114–115, 120–127; "Nogizono zakki," 131, 138–139; "Tokutake-shi jishin kiji," 154–156; "Eikan zasshi," 256–266, 268–269; "Jishin kiji," 269–270; "Chisai satsuyō," 271–272; "Shinshū jishinki," 276–277, 282–283; "Shinano Daijishinki," 287–288, 301–307; and "Sinkōkan," 307–316, all in *DNJS*, vol. 2 (otsu). Especially useful is Usami's diagram of the flooding (*zu* 248-3, *Higai jishin,* 142).

42. "Tokutake-shi jishin kiji," in *DNJS*, vol. 2 (otsu), 156.

43. For an authoritative study of the death and destruction, see Akabane Sadayuki and Inoue Kimio, "Saigai no jōkyō," in Chūō bōsai kaigi, *1847 Zenkōji jishin hōkokusho* (Nihon shisutemu kaihatsu kenkyūsho, 2007), 22–42. See also page 222. Other detailed accounts of the destruction include Shinano mainichi shinbunsha kaihstsukyoku shuppanbu, ed., *Zenkōji daijishin,* 73–108, and "Akahane Sadayuki, 'Zenkōji jishin to saigai no zenbō,'" in Akahane and Kitahara, *Zenkōji jishin,* 9–34.

44. "Koizumi Sōken nichiroku," in *NRJSS,* vol. 2, 227.

45. "Odawara hanshi Hoshimi bō-shokan," in *DNJS,* vol. 2 (*otsu*), 327.

46. Nishimaki Kōzaburō, ed. *Kawaraban shinbun: Edo, Meiji sanbyaku jiken,* vol. 1 (Heibonsha, 1978), 140–141; and Inagaki Fumio, ed., *Edo no taihen, ten no kan* (Heibonsha, 1995), 58. To view this print, see http://gazo.dl.itc .u-tokyo.ac.jp/ishimoto/1/01–021/00001.jpg.

47. Usami, *Higai jishin,* 146–148, and Ishibashi Katsuhiko, *Daijishinran no jidai* (Iwanami shoten, 1994), 8–13.

48. "Jishin kaishō seisetsuroku," in *DNJS,* vol. 2 (*otsu*), 358, 359.

49. Okada, *Jishin chizu,* 146; Sangawa, *Jishin no Nihonshi,* 171–172; Usami, *Higai jishin,* 148–151; and "Wakisaka Antaku nikki" and "Ōsaka jishinki," in *DNJS,* vol. 2 (*otsu*), 330–333, 340–344.

50. Tsuji, "Ansei tōkai, nankai jishin," in Chūō bōsai kaigi, *1854 Ansei tōkai jishin, Ansei nankai jishin hōkokusho,* 2. See also Nishiyama Shōjin, "Ansei nankai jishin ni okeru Ōsaka de no shinsai taiō," in Chūō bōsai kaigi, *1854 Ansei tōkai jishin,* 51, 55–59, 62.

51. Okada, *Jishin chizu,* 119–120; Itō, *Jishin to funka no Nihonshi,* 91–96; Sangawa, *Jishin no Nihonshi,* 173; and Tsuji, "Ansei tōkai, nankai jishin," in Chūō bōsai kaigi, *1854 Ansei tōkai jishin,* 1–2. See also page 132.

52. Okada, *Jishin chizu,* 119; Itō, *Jishin to funka no Nihonshi,* 99–100; Sangawa, *Jishin no Nihonshi,* 180–181; Usami, *Higai jishin,* 151–168; and Tsuji, "Ansei tōkai, nankai jishin" and Kitahara Itoko, "Ansei tōkai, nankai jishin no higai jōhō ni tsuite: kawaraban o chūshin ni," in Chūō bōsai kaigi, *1854 Ansei tōkai jishin,* 84–85.

53. Ishibashi, *Daijishinran,* 25–26.

54. For a detailed table of JMA seismic intensity scale levels, along with important qualifying points, see Chūō bōsai kaigi, *1855 Ansei Edo jishin hōkokusho,* 32–39. See also page 2.

55. For a comprehensive summary of the earthquake, see Usami, *Higai jishin,* 171–182. There are dozens of excellent modern studies of damage patterns, a topic I revisit later. An overall summary in text and in charts is Chūō bōsai kaigi, *1855 Ansei Edo jishin hōkokusho,* 1–11, 25–30 (including charts 1-1 through 1-6). For a systematic collection of all relevant documents and materials, see Sayama Mamoru, *Ansei Edo jishin saigaishi* (hereafter *SGS*).

56. For a tabular summary of these nine documents, see Chūō bōsai kaigi, *1855 Ansei Edo jishin hōkokusho,* Table 1-2, 23–24. Only one document permits even a rough estimate of a specific number. For further discussion of focal depth, see Okada Yoshimitsu, *Saishin Nihon no jishin chizu* (Tōkyō shoseki, 2006), 87; Sangawa Akira, *Jishin no Nihonshi: daichi was nani o kataru ka* (Chūōkōron shinsha, 2007), 191; and Nakamura Misao, Kayano Ichirō, and Matsuura Ritsuko, "Ansei Edo jishin no shubuken de no higai" *Rekishi*

jishin, 19 (2003), 32–33, 36. Nakamura et al. estimate a focal depth of 35–70 kilometers, and many other estimates are within this range. For a detailed seismological discussion of the focal depth of the Ansei Edo earthquake that summarizes many competing theories and approaches, see William H. Bakun, "Magnitude and Location of Historical Earthquakes in Japan and Implications for the 1855 Ansei Edo Earthquake," *Journal of Geophysical Research,* vol. 110, B02304 (2005), 12–22.

57. Okada, *Jishin chizu,* 121–122; Usami, *Higai jishin,* 207–211; and Chūō bōsai kaigi, *1891 Nōbi jishin hōkokusho,* Nihon shisutemu kaihatsu kenkyūsho, 2008, esp. 209.

58. Okada *Jishin chizu* (Tōkyō shoseki, 2006), 52.

59. For an account of modern Tōhoku tsunamigenic earthquakes, see Gregory Smits, "Danger in the Lowground: Historical Context for the March 11, 2011 Tōhoku Earthquake and Tsunami," in *Asia-Pacific Journal* 9, issue 20, no. 4 (May 16, 2011).

60. Itō Kazuaki, *Jishin to funka no Nihonshi* (Iwanami shoten, 2002), 106–107.

61. For a comprehensive survey of the geology and damage from this earthquake, see Usami, *Higai jishin,* 219–230.

62. Koshimura Shun'ichi and Shutō Nobuo, "Meiji Sanriku jishin tsunami," in Chūō bōsai kaigi, *1896 Meiji Sanriku jishin tsunami hōkokusho,* 2005, 15–17.

63. Koshimura and Shutō, "Meiji Sanriku jishin tsunami," 15, 17–20.

64. Okada, *Jishin chizu,* 88–91; Usami, *Higai jishin,* 272–278; and Chūō bōsai kaigi, *1923 Kantō daishinsai hōkokusho, dai ippen* (Nihon shisutemu kaihatsu kenkyūsho, 2006), esp. 238.

65. Okada, *Jishin chizu,* p. 59; Itō, *Jishin to funka no Nihonshi,* 117; and Usami, *Higai jishin,* 302–306. For photos of Ryōri Bay before and after the 1933 tsunami, see http://protea.dbms.cs.gunma-u.ac.jp/TSUNAMI/tsunami_data/IwateShowaShinsai_kawamoto/image/IwateShowaShinsai_00024.jpg.

66. Luke S. Roberts, *Performing the Great Peace: Political Space and Open Secrets in Tokugawa Japan* (Honolulu: University of Hawai'i Press, 2012), esp. 43–52.

67. Mark Ravina, *Land and Lordship in Early Modern Japan* (Stanford, CA: Stanford University Press, 1999).

68. "Chōshin hiroku, jōkan," in *NRJSS,* vol. 3, 212.

69. "Eikan zasshi," in *DNJS,* vol. 2 (*otsu*), 248–249.

70. Regarding the history of the concept of *shinkoku,* see Satō Hiroo, *Shinkoku Nihon* (Chikuma shinsho 591) (Chikuma shobō, 2006) and Kitai Toshio, *Shinkokuron no keifu* (Hōzōkan, 2006).

71. Regarding the significance of Shinano's complex geopolitical circumstances, see Kären Wigen, *A Malleable Map: Geographies of Restoration in Central Japan, 1600–1912* (Berkeley: University of California Press, 2010).

72. On this topic, see Mary Elizabeth Berry, *Japan in Print: Information and Nation in the Early Modern Period* (Berkeley: University of California Press, 2006).

73. Buyō Inshi, *Seji kenmonroku,* Honjō Eijirō, ed. (Seiabō, 2001, originally 1816).

74. For a thorough discussion of this matter, see David L. Howell, *Geographies of Identity in Nineteenth-Century Japan* (Berkeley: University of California Press, 2005), esp. pages 45–78.

75. Fujita Rihei, *Edo kanoko,* Asakura Haruhiko, ed. (Sumiya shobō, 1970, originally 1687), 250–290. See also Jurgis Elisonas, "Notorious Places: A Brief Excursion into the Narrative Topography of Early Edo," in James L. McClain, John W. Merriman, and Ugawa Kaoru, eds., *Edo and Paris: Urban Life and the State in the Early Modern Era* (Ithaca, NY: Cornell University Press, 1994), 284–285; and Berry, *Japan in Print,* 163. See also pages 156–157.

76. Print #90 in Miyata Nobori and Takada Mamoru, eds., *Namazue: Shinsai to Nihon bunka* (Ribun shuppan, 1995), 225, 299–300. To view this print, see http://gazo.dl.itc.u-tokyo.ac.jp/ishimoto/2/02–043/00001.jpg.

77. For a discussion of the importance of free circulation in Tokugawa economic thought, see Mark Metzler and Gregory Smits, "Introduction: The Autonomy of Market Activity and the Emergence of *Keizai* Thought," in Bettina Gramlich-Oka and Gregory Smits, eds., *Economic Thought in Early Modern Japan* (Leiden: Brill, 2010), esp. 12–17. Many of the articles in this volume also discuss the prominence of free circulation as a desirable social and economic situation. See also "Ansei itsubō jishin kibun," in *NJS,* vol. 5, supplement 2, part 1, 460 for thoughts on those who hinder the circulation of gold and silver in society.

78. For a trenchant analysis of these matters, see Hayashi, Reiko, "Provisioning Edo in the Early Eighteenth Century: The Pricing Policies of the Shogunate and the Crisis of 1733," in James L. McClain, John W. Merriman, and Ugawa Kaoru, eds. *Edo and Paris: Urban Life and the State in the Early Modern Era* (Ithaca, NY: Cornell University Press), 1994, 211–233.

79. Takeuchi Makoto, "Festivals and Fights: The Law and the People of Edo," in McClain, Merriman, and Ugawa, *Edo and Paris,* 404.

80. Peter Kornicki, *The Book in Japan: A Cultural History from the Beginning to the Nineteenth Century* (Leiden: Brill, 1998), 350.

81. Katō Takashi, "Governing Edo," in McClain, Merriman, and Ugawa, *Edo and Paris,* 53.

82. For a study of litigiousness in a rural area, see Herman Ooms, *Tokugawa Village Practice: Class, Status, Power, Law* (Berkeley: University of California Press, 1996).

83. Takeuchi, "Festivals and Fights," 384.

84. Sasaki Junnosuke, *Yonaoshi* (Iwanami shoten, 1979), 12–15.

85. Herbert P. Bix, *Peasant Protest in Japan, 1590–1884* (New Haven, CT: Yale University Press, 1986), 145.

86. Asai Ryōi, *Kaname'ishi* (1662), in Taniwaki Masachika, Oka Masahiko, and Inoue Kazuhito, eds. and trans., *Kanazōshishū* (Shōgakkan, 1999), 15. For an 1830 example, see *DNJS*, vol. 1 (*kō*), 567. See also Kitani, *Namazue no shinkō,* 41, and Nishiyama Akihito, "Kyōto de no higai to jishin taiō," in Chūō bōsai kaigi, *1662 Kanbun Ōmi-Wakasa jishin,* 141.

87. "Daijishin rokka nukigaki kyōka," in *NRJSS*, vol. 3, 288.

88. Kitahara Itoko, *Jishin no shakaishi: Ansei daijishin to minshū* (Kōdansha, 2000), 98–99, and Noguchi, *Ansei Edo jishin,* 201–202.

89. Kitahara Itoko, "Saigai to kawaraban: Sono rekishiteki tenkai," in Kinoshita Naoyuki and Yoshimi Shunya, eds., *Nyūsu no tanjō: Kawaraban to shinbun nishikie no jōhōsekai* (Tōkyō daigaku sōgō kenkyū hakubutsukan, 1999), 25–26.

90. Kornicki, *The Book in Japan,* 335.

91. Ibid., 351.

92. Berry, *Japan in Print,* 107–111.

93. Shimizu Isao, *Edo no manga* (Kōdansha, 2003), 131–132.

94. Richard Rubinger, *Popular Literacy in Early Modern Japan* (Honolulu: University of Hawai'i Press, 2007), 160.

95. Kornicki, *The Book in Japan,* 275–276.

96. Henry D. Smith II, "The History of the Book in Edo and Paris," in McClain, Merriman, and Ugawa, *Edo and Paris,* 336, 348. See also Gerald Groemer, "Singing the News: Yomiuri in Japan during the Edo and Meiji Periods," *Harvard Journal of Asiatic Studies* 54, no. 1 (June 1994): 233–261.

97. See Satō, *Shinkoku Nihon,* and Kitai, *Shinkokuron* for thorough discussions of this topic.

98. Noguchi, *Ansei Edo jishin,* 195–196.

99. The major study of this phenomenon is Miyata Noboru, *Kinsei no hayarigami* (Hyōronsha, 1972).

100. As examples, see the prints *Manzairaku mi no yōjin* (Joyous self-precaution), #69 in Miyata and Takada, *Namazue,* 285–286 (to view this print, see http://gazo.dl.itc.u-tokyo.ac.jp/ishimoto/2/02-047/00001.jpg; *Namazu kakejiku* (Catfish hanging scroll), #129 in Miyata and Takada, *Namazue,* 222, 320 (to view this print, see http://dl.ndl.go.jp/info:ndljp/pid/1302032); and *Jishin myōsaku kudokusan* (The wondrous efficacy of the earthquake's blessing), #196 in Miyata and Takada, *Namazue,* 357 (to view this print, see http://gazo.dl.itc.u-tokyo.ac.jp/ishimoto/2/02-039/00001.jpg).

101. Sarah E. Thompson, for example, states, "There was also the possibility for implied criticism of the government in the reporting of current events,

especially given the ancient notion, imported from China, that a truly virtuous regime would be so completely uneventful that even natural disasters would not occur. The suppression of news reporting may have been due in part to a desire to suggest this ideal condition." "The Politics of Japanese Prints," in Sarah E. Thompson and H. D. Harootunian, *Undercurrents of the Floating World: Censorship and Japanese Prints* (New York: Asia Society Galleries, 1991), 34.

102. Earthquake-induced era name changes were Tengyō (938), Ten'en (973), Jōgen (976), Eichō (1096), Bunji (1185), Keichō (1596), Tenpō (1830), and Ansei (1854). See Yamamoto Takeo, "Shiryō ginmi no hitsuyōsei," in Hagiwara Takahiro et al., *Kojishin: rekishi shiryō to katsu-dansō kara saguru* (Tōkyō daigaku shuppankai, 1982), 46. To Yamamoto's list we could add Hōei (1704), owing to the December 31, 1703, Genroku earthquake.

103. Quoted in Harold Bolitho, "The Tempō Crisis," in Marius B. Jansen, ed., *The Cambridge History of Japan,* vol. 5, *The Nineteenth Century* (New York: Cambridge University Press, 1989), 117.

104. Noguchi, *Ansei Edo jishin,* 39.

105. See Tahara Tsugio, ed., *Yamaga Sokō,* Nihon no meicho 12 (Chūō kōronsha, 1983), 139; Miura Baien, *Zeigo,* in Yamada Keiji, ed., trans. *Miura Baien,* Nihon no meicho 20 (Chūō kōronsha, 1984), 482–483; and (regarding Motoori Norinaga) Tahara Tsugio and Morimoto Jun'ichirō, eds., *Yamaga Sokō,* Nihon shisō taikei 32 (Iwanami shoten, 1970), 334, 542–543.

106. Groemer, "Singing the News," 245.

107. "Jishin no ben" in the print *Shokoku daijishin,* figure 19 in Kitahara, "Saigai to Kawaraban," 31. For the web version, see http://www.um.u-tokyo.ac.jp/publish_db/1999news/02/20203.html (accessed January 20, 2012).

108. Most mainstream seismologists regard earthquake prediction in the narrow sense of specifying location, specific time, and magnitude as impossible. Nevertheless, there is no shortage of publications claiming, in retrospect, that some past major earthquake could have been predicted if we had only been better attuned to nature's signals. Of the seven chapters in a book inspired by the Indian Ocean tsunami of 2004, one argues that statistical analysis proves there are nonrandom components to the spatial and temporal distribution of earthquakes. The other six chapters posit specific precursors: the gravitational pull of the sun at certain times of the year; earthquake vapor and clouds; sunspot cycles; abnormally high temperatures combined with astro-tidal triggering; star storms; and sunspots. Although wrapped in a veneer of modern science, the basic explanation for all of these alleged precursors would have made sense to educated early modern Japanese. See Saumitra Mukherjee, ed., *Earthquake Prediction* (Leiden: Brill, 2006).

109. Jelle Zeilinga de Boer and Donald Theodore Sanders, *Earthquakes in Human History: The Far-Reaching Effects of Seismic Disruptions* (Princeton, NJ: Princeton University Press, 2005), 172–173.

110. Hidemi Shiga, "The Catfish Underground: Japan's Earthquake Folklore and Popular Responses to Disaster," *Orientations Magazine for Collectors and Connoisseurs of Asian Art* (April 2006), 77.

111. Tsuji Yoshinobu, *Zukai, Naze okoru? Itsu okoru? Jishin no mekanizumu* (Nagaoka shoten, 2010), 206.

112. Gregory Clancey, *Earthquake Nation: The Cultural Politics of Japanese Seismicity, 1868–1930* (Berkeley: University of California Press, 2006), 216. I should point out that more recently Clancey has characterized the earthquake catfish as a metaphor, not a literal belief (e.g., presentation given at the symposium "Thinking through Disasters: Japanese Earthquakes Past and Present," Columbia University, April 6, 2012).

113. Musha, *Jishin namazu,* 12–22.

114. "Sagami no kuni ōjishin," in Inagaki Fumio, ed., *Edo no taihen, ten no kan* (Heibonsha, 1995), 58. See also Nishimaki Kōzaburō, ed., *Kawaraban shinbun: Edo, Meiji sanbyaku jiken,* vol. 1 (Heibonsha, 1978), 140–141, and "Saigai to kawaraban: Sono rekishiteki tenkai," in Kinoshita and Yoshimi, *Nyūsu no tanjō,* 32–33. To view this print, see http://gazo.dl.itc.u-tokyo.ac.jp/ishimoto/1/01–021/00001.jpg.

115. Print #84 in Miyata and Takada, *Namazue,* 297. To view this print see http://dl.ndl.go.jp/info:ndljp/pid/1302131.

116. Tomisawa Tatsuzō, "'Namazue no sekai' to minzoku ishiki," *Nihon minzokugaku,* no. 207 (1996), 102.

117. Print #18 in Miyata and Takada, eds. *Namazue,* 248. To view this image, see http://gazo.dl.itc.u-tokyo.ac.jp/ishimoto/1/01–043/00001.jpg.

118. *Kojiruien* (Dictionary of Historical Terms) Database (Kyoto: Nichibunken, International Research Center for Japanese Studies, 2007), 1356–1357, in "Chibu," entry "Jishin," http://www.nichibun.ac.jp/graphicversion/dbase/kojirui_e.html (accessed December 21, 2011). See also Hashimoto, *Jishingaku,* 19.

119. Quoted in Hashimoto, *Jishingaku,* 20.

Chapter 2   Why the Earth Shakes

1. How Ryōi can make this claim is unclear, because at the beginning of *Foundation Stone* he clearly states that the earthquake began on the first day of the fifth month, and all other sources are in agreement with this date. There was a rich tradition in Japanese and Chinese lore of earthquakes presaging other events—often outbreaks of disease, famine, military problems, and so on. For more details, see Hagiwara Takahiro et al., *Kojishin: Rekishi shiryō to katsudansō kara saguru* (Tōkyō daigaku shuppankai, 1982), 45–46; Unno Kazutaka,

"Kigan, majinai no tsukawareta Nihonzu," in Unno Kazutaka, *Tōyō chirigakushi kenkyū, Nihon hen* (Osaka: Seibundō shuppan, 2005), 222, 251 (n. 15); and Hashimoto Manpei *Jishingaku kotohajime: kaitakusha Sekiya Seikei no shōgai* (Asahi shinbunsha, 1983), 49–54. Relevant to Ryōi's use of Su Dongpo's poem, Hashimoto (p. 50) and Unno explain that in Buddhist-influenced earthquake divination, earthquakes might be caused by the movement of the fire (or hearth) deity, the dragon deity, the golden-winged bird, or the celestial king. Depending on which month or days of the month these types of earthquakes occur, they could portend either good or bad fortune.

2. Asai Ryōi, *Kaname'ishi* (1662), in Taniwaki Masachika, Oka Masahiko, and Inoue Kazuhito, eds., trans., *Kanazōshishū* (Shōgakkan, 1999), 82–83. For details on the history of Kashima's Foundation Stone in connection with earthquakes, see Gregory Smits, "Conduits of Power: What the Origins of Japan's Earthquake Catfish Reveal about Religious Geography," *Japan Review* 24 (2012): 41–65.

3. Early accounts of *kami* often described them as inscrutable, allotting curses (*tatari*) for no discernible reason. Medieval *kami,* by contrast, usually doled out rewards and punishments in accordance with and as exemplifiers of the norms of human society. See Satō Hiroo, *Shinkoku Nihon,* Chikuma shinsho 591 (Chikuma shobō, 2006), esp. 67–71.

4. This basic concept derives from Chinese thought and closely resembles Aristotelian notions of earthquake mechanics in which underground pneuma is the substance that causes the earth to shake. As Emanuela Guidoboni and John E. Ebel point out, "In China, as in Europe, the pneumatic theory survived for a long time and was extraordinarily popular for almost two thousand years, becoming the most enduring and widespread theory that has ever been developed." Guidoboni and Ebel, *Earthquakes and Tsunamis in the Past: A Guide to Techniques in Historical Seismology* (New York: Cambridge University Press, 2009), 152.

5. Mizuno Shōji, "Chūsei no saigaikan," in Kitahara Itoko, ed., *Nihon saigaishi* (Yoshikawa kōbunkan, 2006), 97–98. For a detailed discussion of Chinese-derived *tenjin-sōkan* (heaven and human interconnection) thought in the context of earthquakes in Japan, see Hagiwara et al., *Kojishin,* 43–47. Regarding countermeasures, see 44–45, 46–47. For a discussion of era name changes because of earthquakes, see 46–47 and Nihon gakushiin, eds., *Meiji-zen Nihon butsuri kagakushi* (Nihon gakjutsu shinkōkai, 1964), 501.

6. Mizuno, "Chūsei," 98–99, and Imahori Ta'itsu, "Kokudo no saigai to akukishin: Saigai to zokushin," in Inseiki bunka kenkyūkai, eds., *Seikatsushi* (Inseiki bunka ronshū, vol. 5), 198–232. According to the *Niō Sutra,* for example, "Disorder among the deities leads to disorder in a country. Because the deities are in disorder, so too are the masses of people" (Imahori, "Kokudo no saigai," 224).

7. Philip B. Yampolsky, ed., Burton Watson et al., trans., *Selected Writings of Nichiren* (New York: Columbia University Press, 1990), 16. For the full discussion, see 13–21. See also Mizuno, "Chūsei," 99, and Imahori, "Kokudo no saigai," 220–226.

8. *Azuma kagami,* Kangi 3, 5m 17d. "To attain peace in the realm and a bountiful harvest for the country, from today, at the Tsuruoka Hachiman Shrine, thirty priests will intone the *Dai hannyakyō.*" (http://www5a.biglobe .ne.jp/~micro-8/toshio/azuma/123105.html, accessed February 14, 2011). See also Mizumo, "Chūsei," 100–101, and Hagiwara et al., *Kojishin,* 46.

9. Herman Ooms, *Imperial Politics and Symbolics in Ancient Japan: The Tenmu Dynasty, 650–800* (Honolulu: University of Hawai'i Press, 2009), 90.

10. Kitahara Itoko, ed., *Nihon Saigaishi* (Yoshikawa kōbunkan, 2006), 161, and Ueda Kazue and Usami Tatsuo, "Yūshi irai no jishin kaisū no hensen," *Rekishi jishin* 6 (1990): 181–187.

11. *Taikyoku jishinki,* in Aoki Kunio et al., eds., *Taikyoku jishinki, Ansei kenbunroku, Jishin yobōsetsu, Bōkasaku zusetsu,* Edo kagaku koten sōsho, vol. 19 (Kōwa shuppan, 1979), 12. See also Hashimoto, *Jishingaku,* 11–12, and Nihon gakushiin, *Butsuri kagakushi,* 534–538.

12. *Taikyoku jishinki,* 12.

13. Ibid., 15.

14. Ibid., 17.

15. Ibid., 18.

16. Ibid., 19.

17. Ibid., 20.

18. Asai Ryōi, *Kaname'ishi,* 66–70.

19. Hashimoto, *Jishingaku,* 7–8, and Guidoboni and Ebel, *Historical Seismology,* 148–149.

20. Quoted in Nihon gakushiin, *Butsuri kagakushi,* 533. Regarding *Kenkon bensetsu,* see 532–524 and Hashimoto, *Jishingaku,* 13.

21. Nihon gakushiin, *Butsuri kagakushi,* 533, and Hashimoto, *Jishingaku,* 13. According to Hashimoto, these *Kenkon bensetsu* explanations are modifications of Aristotle's original theory.

22. Miura Baien, *Zeigo,* in Yamada Keiji, ed., trans., *Miura Baien,* Nihon no meicho 20 (Chūō kōronsha, 1984), 454.

23. Baba Nobutake, for example, used language almost identical to *Tenkei wakumon* in his 1706 *Shogaku tenmon shinan.* See Hashimoto, *Jishingaku,* 14–15.

24. Although gunpowder served as a metaphor in *Tenkei wakumon,* several later Japanese scholars of Dutch studies proposed actual gunpowder as a possible cause of earthquakes. For example, Kawamoto Kōmin's *Kikaikanran kōgi,* published soon after the Ansei Edo earthquake, proposed two theories of underground combustion, one of which was that heated oxygen ignites

saltpeter, charcoal, and sulfur located near the focus of an earthquake. Another book published after the Ansei Edo earthquake was Hirose Genkyō's *Rigaku teiyō*, based on his reading of several Dutch books. He proposed that saltpeter, charcoal, and sulfur under the earth explode to create earthquakes. See Hashimoto, *Jishingaku*, 36–38, and Nihon gakushiin, *Butsuri kagakushi*, 552–553.

25. *Tenkei Wakumon*, vol. 2, "Jishin" (Osaka: Ōsaka shobō, 1794 revision of the 1730 edition), manuscript, 28–29 (four pages). See also Hashimoto, *Jishingaku*, 13.

26. Sima Qian, *Shijing*, Zhoubenji, King You 2, entry 36. William H. Nienhauser Jr., ed., Tsai-fa Cheng et al., trans., *The Grand Scribe's Records*, vol. 1: *The Basic Annals of Pre-Han China by Ssu-ma Ch'ien* (Bloomington: Indiana University Press, 1994), 73. For the original text, see http://ctext.org/shiji/zhou-ben-ji.

27. Terashima Ryōan, *Wakan sansai zue*, vol. 8, Shimada Isao, Takeshima Atsuo, and Higuchi Motomi, trans., eds. (Heibonsha, 1987), 9.

28. Terashima, *Wakan sansai zue*, 9–10.

29. Terashima, *Wakan sansai zue*, 11. Although even today it is common for many people to assume a connection between earthquakes and volcanic activity, scientists have not been able to establish any direct link. For a basic explanation, see Shimamura Hideki, *Nihonjin ga shiritai jishin no gimon rokujūroku: Jishin ga ōi Nihon dakara koso chishiki no sonae mo wasurezu ni* (Sofutobanku kurieitibu, 2008), 90–92.

30. Terashima, *Wakan sansai zue*, 12.

31. Quoted in Nakamura Misao, "Ansei Edo jishin," in Chūō bōsai kaigi, *1855 Ansei Edo jishin hōkokusho* (Fuji sōgō kenkyūsho, 2004), 6. Nakamura points out that by 1855, experience from other earthquakes had promoted an understanding of soil base as a major factor in the intensity of shaking. See also "Yabure mado no ki," in *DNJS*, vol. 2 (*otsu*), 561.

32. "Nai no nochimigusa," in *DNJS*, vol. 2 (*otsu*), 572.

33. See, for example, Susan Elizabeth Hough, *Earthshaking Science: What We Know (and Don't Know) about Earthquakes* (Princeton, NJ: Princeton University Press, 2002), 80–83.

34. Terashima, *Wakan sansai zue*, 12.

35. "Chika yori kaki o hassuru no jō," in Hattori Yasunari (text) and Utagawa Yoshiharu (illustrations), *Ansei kenmonroku* (hereafter *AKR*), vol. 2, 12. See also Arakawa, *Jitsuroku, Ō-Edo kaimetsu no hi*, 89.

36. "Kokuryōki," in *DNJS*, vol. 1 (*kō*) (Shibunkaku, 1973), 324. In this case, ships unable to dock at Tsuro and Murozu were able to dock at Kōchi for the first time because of approximately two meters of subsidence in the area. See Okada Yoshimitsu, *Saishin Nihon no jishin chizu* (Tōkyō shoseki, 2006), 169–170.

37. Nishikawa Joken, *Kaii bendan,* vol. 5 (Kyoto: Ryūshiken, 1715). First four page faces after the heading "Jishin" deal with China; next two-plus page faces deal with Japan.

38. Nishikawa, *Kaii bendan,* vol. 5, page faces 8–12 after the heading "Jishin." See also Nihon gakushiin, *Butsuri kagakushi,* 538–539.

39. For a detailed examination of this and related matters, see Smits, "Conduits of Power," 41–65.

40. Nishikawa, *Kaii bendan,* vol. 5, page faces 12–14 after the heading "Jishin." See also Nihon gakushiin, *Butsuri kagakushi,* 539. The later transcription spells the earthquake-prone country as "Beruru," but it is clearly "Berū" in the original.

41. Nishikawa, *Kaii bendan,* vol. 5, page faces 1–2 after the heading "Chiretsu oyobi ni yama-sakuru."

42. Nishikawa, *Kaii bendan,* vol. 5, page face 3 after the heading "Chiretsu oyobi ni yama-sakuru." See also Nihon gakushiin, *Butsuri kagakushi,* 539.

43. Nishikawa, *Kaii bendan,* vol. 5.

44. Nihon gakushiin, *Butsuri kagakushi,* 540–550.

45. Miura, *Zeigo,* 421.

46. Kojima Tōzan and Tōrōan-shujin, *Jishinkō* (Kyoto: Saiseikan, 1830), first two page faces in the second section of the book, and Hashimoto, *Jishingaku,* 22–24.

47. Kojima, *Jishinkō,* page faces 1–3 after the heading "Jishinkō" describe the 1830 earthquake and printed page faces 3–8 describe past earthquakes. "Jishinkō," in *DNJS,* vol. 1 (*kō*), 589–590. See also Hashimoto, *Jishingaku,* 23–24, and Nihon gakushiin, *Butsuri kagakushi,* 562–563.

48. Kojima, *Jishinkō,* page faces 1–4 after the heading "Jishin no setsu." "Jishinkō," in *DNJS,* vol. 1 (*kō*), 590.

49. See, for example, Unno Kazutaka, "Maps of Japan Used in Prayer Rites or as Charms," *Imago Mundi* 46 (1994), note on 76.

50. Kojima, *Jishinkō,* final two page faces in the section "Jishin no shirushi." A *jishin mushi* illustration appears the second part of the book. "Jishinkō," in *DNJS,* vol. 1 (*kō*), 591 (no illustrations). See also Hashimoto, *Jishingaku,* 24. To view the illustration, see figure 2.

51. Kojima, *Jishinkō,* page faces 1–2 in the second section. "Jishinkō," in *DNJS,* vol. 1 (*kō*), 591. See also Hashimoto, *Jishingaku,* 24.

52. That the earth is a sphere was common knowledge throughout the Tokugawa period. Ancient Greek mathematicians proved that the earth was a sphere, and around 240 BCE, Eratosthenes calculated the circumference of the earth to a high degree of accuracy. The idea of a spherical earth later spread to India and the Islamic world, where scholars quickly accepted it. In China the idea did not become firmly established until the Ming dynasty, in part owing

to the work of Jesuit priests. In Japan, there were explicit debates about the shape of the earth and heavens during the first decade of the seventeenth century. Kyoto was a center for astronomy, and Carlo Spinola established a mathematics and astronomy academy there in 1611. See Sugimoto Isao, ed., *Kagakushi,* Taikei Nihonshi sōsho 19 (Yamakawa shuppan, 1976), 136.

53. In most accounts, earthquakes simply occur, but sometimes they arrive from a particular direction. For example, from within Kyoto, Asai Ryōi described the Kanbun earthquake as arriving from the northeast. See *Kaname'ishi,* 15. A diary describing the 1847 Zenkōji earthquake stated, "An earthquake arrived from the west." See "Koizumi Sōken nichiroku," in *NRJSS,* vol. 2, 226.

54. Kojima, *Jishinkō,* page faces 2–5 in the second section. The third page face consists of an illustration of the earthquake center. "Jishinkō," in *DNJS,* vol. 1 (*kō*), 592 (no illustration). See also Hashimoto, *Jishingaku,* 24–25, and Nihon gakushiin, *Butsuri kagakushi,* 563. To view the earthquake center illustration, see figure 3.

55. See "Jishin no hōkaku o iu jō," in *AKR,* vol. 2, 17–18 (final three page faces). See also Arakawa, *Jitsuroku, Ō-Edo kaimetsu no hi,* 69–71. To view the diagram, see http://archive.wul.waseda.ac.jp/kosho/wo01/wo01_03628/wo01_03628_0002/wo01_03628_0002_p0022.jpg.

56. Kojima, *Jishinkō,* page faces 10–12 in the second section. "Jishinkō," in *DNJS,* vol. 1 (*kō*), 593 (no illustration).

57. Kojima, *Jishinkō,* page face 16 in the second section.

58. Ibid., section entitled "Daijishinkō no ato."

59. Ibid., page face 3 in the second section. To view the diagram, see figure 3.

60. In examining the broad discourse on earthquakes, which includes hundreds of minor works, one can occasionally find variations on this idea. For example, one minor text from 1855 or 1856 explains that yin rising upward from out of the earth becomes rain, clouds, or in extreme cases, thunder and lightning. Conversely, yang entering the earth from the sun, in extreme concentrations, causes earthquakes. See "Ansei ni Edo jishin," in *NJS,* vol. 5, supplement 2 (*Ansei Edo jishin*), part 1, 575.

61. Toyo Tokinari, *Honchō jishinki,* in Edo josei bunko, vol. 49 (Ōzorasha, 1994), no pagination. See the final page face. The discussion of the causes and warning signs of earthquakes can be found in *Kojiruien* (an encyclopedic work published between 1896 and 1914), 1359 in "Chibu" (available today via the Kojiruien Database, International Research Center for Japanese Studies: http://shinku.nichibun.ac.jp/kojiruien/). See also Hashimoto, *Jishingaku,* 26–27.

62. Toyo, *Honchō jishinki,* first page face of the main text, and *Kojiruien,* 1359 in "Chibu," entry "Jishin." This same description occurs in the similarly titled 1855 *Honchō jishin no shidai.* See *DNJS,* vol. 2 (*otsu*), 584 and *FN,* 563.

63. Toyo, *Honchō jishinki,* page faces 2–6 from the start of the main text, and *Kojiruien,* 1359 in "Chibu," entry "Jishin."

64. Toyo, *Honchō jishinki.* Page faces 7–25 from the start of the main text (including the illustrations) constitute the survey of earthquakes up to 1830. See also the comprehensive listing of major historical earthquakes in the "Jishin" section of *Kojiruien* under the subheading "Jishin rei," 1366–1375.

65. Hashimoto, *Jishingaku,* 26.

66. "Denchū nikki," in *DNJS,* vol. 1 (*kō*), 247, 248. See also "Fusō Nikki," in *NJS, hoi* (supplement), 148. In Kyoto, "fearing earthquakes [= aftershocks], people lived in huts they erected outside."

67. "Keien ryakki," in *DNJS,* vol. 1 (*kō*), 249.

68. *Konoe nikki,* for example, records aftershocks for many days in Kyoto, the most common phrase being *tokidoki jishin* (earthquakes [= aftershocks] from time to time). The last day on which an aftershock is recorded was 1662/12/7. See *NTJSS,* 149–154. The *Matsuo-ke ruidai nikki* (Diary of the Matsuo house generations) typically employs the phrases *jishin tokidoki, jishin shō tokidoki* (a few earthquakes [= aftershocks] from time to time), or *jishin shō* (a few earthquakes [= aftershocks]). Its last recording of an aftershock was 1662/8/9. See *NTJSS,* 154–156.

69. "Mitsuhisa Kyūki," in *DNJS,* vol. 1 (*kō*), 533–535.

70. "Kyōto jishin kenmonki," in *DNJS,* vol. 1 (*kō*), 569.

71. "Ibarada Gunshi kenmon satsuyō hikae," in *NRJSS,* vol. 2, 212.

72. Quoted in Miki, *Kyōto daijishin,* 111.

73. *Ansei kenmonshi* (illustrations by Utagawa Kuniyoshi et al.) (hereafter *AKS*), vol. 1, fifth page face of main text (including illustrations). See also Arakawa Hidetoshi, ed., *Jitsuroku, Ō-Edo kaimetsu no hi: Ansei kenmonroku, Ansei kenmonshi, Ansei fūbunshū* (Kyōikusha, 1982), 112. To view this image, see figure 4.

74. "Yabure mado no ki," in *DNJS,* vol. 2 (*otsu*), 566 (a similar but less detailed system of notation is found in "Honchō jishin no shidai," 585–586), and "Fujiokaya nikki," in *NJS,* vol. 5, supplement 2, part 1, 330.

75. Hashimoto, *Jishingaku,* 33. This matter is reported in *AKR,* vol. 3, 10 (*ge no jū*). See also Arakawa, *Jitsuroku, Ō-Edo kaimetsu no hi,* 85.

76. "Kamahara Dōzan jishin kiji," in *DNJS,* vol. 2 (*otsu*), 86–88. Apparently there was another earthquake on the twenty-second day of the tenth month, and aftershocks continued for successive days thereafter. Several of these aftershocks were as large as the initial quake, so that second main shock must have been mild relative to the main shock of the Zenkōji earthquake. See also Shinano mainichi shinbunsha kaihstsukyoku shuppanbu, ed., *Zenkōji daijishin,* 100–103.

77. Asai, *Kaname'ishi,* 13–39. See also Kitahara, "Saigai to jōhō," 236.

78. *NJS, hoi* (supplement), 149. The rounded, substantial part is labeled "front," and the part resembling a handle or tail is labeled "back." The text explains that this object emerged from a giant star, flew across the sky from the southwest (*hitsujisaru*) to the northeast (*ushitora*), and disappeared. Clearly, it was not seen as a comet, but a comet may have been the only frame of reference for conceiving of the object's shape.

79. For a summary of earthquake lights, see Susan Hough, *Predicting the Unpredictable: The Tumultuous Science of Earthquake Prediction* (Princeton, NJ: Princeton University Press, 2010), 135.

80. "Kanrosō," in *DNJS*, vol. 1 (*kō*), 287. See also Musha Kinkichi, *Jishin namazu* (Meiseki shoten, 1995, originally published in 1957), 53. Musha, who took this and other accounts of light flashes seriously as an integral component or concomitant of earthquakes, proposed that these light flashes were the result of a series of aftershocks.

81. Anonymous, *Jishin narabi ni shikka saikenki* (publisher unknown, 1855), 8–9 (text spanning the fifteenth and sixteenth page faces). See also Musha, *Jishin namazu,* 58.

82. "Jifūroku," in *DNJS*, vol. 2 (*otsu*), 537.

83. Ōsaki Shōji, ed., *Kinsei Nihon tenmon shiryō* (Hara shobō, 1994), 408. I thank Laura Nenzi for this reference.

84. "Chika yori kaki o hassuru no jō," in *AKR*, vol. 3, 11. See also Arakawa, *Jitsuroku, Ō-Edo kaimetsu no hi,* 88, and Musha, *Jishin namazu,* 58.

85. For example, "Ansei itsubō jishin kibun," in *NJS*, vol. 5, supplement 2, part 1, 434.

86. Anonymous, *Edo Ōjishin matsudai hanashi no tane,* 1855, 11. To view the image, see http://archive.wul.waseda.ac.jp/kosho/wo01/wo01_03639/wo01_03639_p0012.jpg. See also Musha, *Jishin namazu,* 59, and Abe Yasunari, "Jishin to hitobito no sōzōryoku," in Chūō bōsai kaigi, *1855 Ansei Edo jishin hōkokusho* (Fuji sōgō kenkyūjo, 2004), 131.

87. *FN,* 516. For a complete listing of these and many other documents from this earthquake dealing with emissions of light and clouds, see *SGS*, vol. 2 (*ge*), 951–955.

88. *AKS*, vol. 2, illustration and text as insert between 10 and 11. See also Arakawa, *Jitsuroku, Ō-Edo kaimetsu no hi,* 152. The *Kenmonshi* account simply reports the damaged spire in the context of damage to other shrines and temples. To view this image, see http://archive.wul.waseda.ac.jp/kosho/wo01/wo01_04209/wo01_04209_0002/wo01_04209_0002_p0012.jpg. For other examples of the damaged spire in popular prints and books, see Nakayama Einosuke, "Ansei no kyodai jishin to kawaraban," in Inagaki Funio, ed., *Edo no taihen: Jishin, kaminari, kaji, kaibutsu* (Heibonsha, 1995), 76–77. For "catfish" print examples (without catfish), see #6 and #22 in Miyata

Noboru and Takada Mamoru, eds., *Namazue: shinsai to Nihon bunka* (Ribun shuppan, 1995), 138, 241–242, 251–252. Print #6 does not mention the beam of ki or light but instead attributes the bending to deterioration of the spire's core support, which had not been replaced in 136 years. To view print #6, see http://gazo.dl.itc.u-tokyo.ac.jp/ishimoto/2/02–038/00001.jpg. For the Meiji example, see "Jishin to arashi," in *Yomiuri shinbun,* January 4, 1885, special edition, 3.

89. Yoshimura Akira, *Sanriku kaigan ōtsunami* (Bungei shunjū, 2004), 82–87.

90. Musha, *Jishin namazu,* 52–104. Musha argues that accounts of flashes of light, pillars of fire, and so on "should not be carelessly dismissed as absurd. It is hardly the case that people of the past purposely wrote lies. . . . Accounts of things such as earthquake light could not have been written by the imagination" (52). He then goes on to analyze the matter for over fifty pages. Musha's earliest example is from 1257, and the next one is from 1703. He seems not to have been aware of *Foundation Stone* and other Kanbun earthquake materials.

91. Miki Haruo, *Kyōto daijishin* (Shibunkaku shuppan, 1979), 60–66.

92. Drawing on the work of psychologist Kikuchi Satoshi, Shimamura explains that the common phenomenon of the discovery of precursors after the fact of an earthquake is an example of "erroneous correlation" (*sakugo sōkan*), whereby preconceived notions combined with the shock of a major earthquake can result in the recall, or manufacture, of precursors from events vaguely remembered. See Shimamura, *Jishin no gimon,* 102–103.

93. "Chōshin hiroku, jōkan," in *NRJSS,* vol. 3, 212–218. The passing reference to "Western learning" may refer to notions of explosions under the earth caused by electricity, gunpowder, or other combustible materials. See chapter 6 for further discussion.

94. Kojima, *Jishinkō,* first page face in the section "Jishin no shirushi." "Jishinkō," in *DNJS,* vol. 1 (*kō*), 590–591. See also Hashimoto, *Jishingaku,* 24, and Nihon gakushiin, *Butsuri kagakushi,* 563.

95. Kojima, *Jishinkō,* page faces 6–10 in the second section. "Jishinkō," in *DNJS,* vol. 1 (*kō*), 592. See also Hashimoto, *Jishingaku,* 25, Nihon gakushiin, *Butsuri kagakushi,* 564–565, and Miki, *Kyōto daijishin,* 60–66.

96. *FN,* 531 and *SGS,* vol. 2 (*ge*), 951–952.

97. *FN,* 556.

98. Ikeya, Motoji, *Earthquakes and Animals: From Folk Legends to Science* (River Edge, NJ: World Scientific, 2004), 15–16, and Chihiro Yamanaka, Hiroshi Asahara, Yutaka Emoto, and Yuko Esaki, "Earthquake Precursors: From Legends to Science and a Possible Early Warning System," in Ülkü Ulusoy and Himansu Kumar Kundu, eds., *Future Systems for Earthquake Early Warning* (New York: Nova Science Publishers, 2008), 204, 205.

99. *FN,* 556, and *AKS,* vol. 3, 19–20. To view this device, see http://archive.wul
    .waseda.ac.jp/kosho/wo01/wo01_04209/wo01_04209_0003/wo01_04209_
    0003_p0019.jpg. See also Arakawa, *Jitsuroku, Ō-Edo kaimetsu no hi,* 191–192;
    Hashimoto, *Jishingaku,* 30–31; Usami, "Kaisetsu," 43–45; Clancey, *Earthquake
    Nation,* 153; and Tsuji Yoshinobu, *Sennen shinsai: Kurikaesu jishin to tsunami
    no rekishi ni manabu* (Daiyamondo sha, 2011), 93–95.

100. According to Ikeya, the loss of magnetism in the stone "led to the immediate
     construction of an earthquake prediction apparatus." He claims to have
     constructed a working replica in his laboratory based on the drawing in *Ansei
     kenmonshi.* Ikeya, *Earthquakes and Animals,* 15. See also Yamanaka et al.,
     "Earthquake Precursors," 204, 205.

101. Tōkyō kagaku hakubutsukan, ed., *Edo jidai no kagaku* (Meicho kankōkai,
     1980, originally published 1934), 257–258.

102. Murayama Masataka, *Shinden kōsetsu* (1856), in *Edo josei bunko,* vol. 49
     (Ōzorasha, 1994, no pagination). See also Tōkyō kagaku hakubutsukan, *Edo
     Jidai no kagaku,* 268.

103. Hashimoto, *Jishingaku,* 20–21.

104. *AKS,* vol. 1, approximately full page 15 of text after the table of contents. See
     also Arakawa, *Jitsuroku, Ō-Edo kaimetsu no hi,* 121.

105. Miyata Noboru, "Toshi minzokugaku kara mita namazu shinkō," in Miyata
     and Takada, *Namazue,* 24–33, and Miyata Noboru, *Kinsei no hayarigami*
     (Hyōronsha, 1972), 158–164.

106. Saitō Gesshin, *Bukō nenpyō 1,* Tōyō bunko #116, Kaneko Mitsuharu, comp.,
     ed. (Heibonsha, 1968, 1992), 103.

107. Hata Ginkei, *Shigure no sode,* part 2 (*kōhen*), vol. 1, in Edo sōsho kankōkai,
     ed. *Edo sōsho,* vol. 10 (Meicho kankōkai, 1961), 107–122.

108. Hata, *Shigure no sode,* 110.

109. *AKR,* vol. 1, 1 (*jō no ichi*). See also Arakawa, *Jitsuroku, Ō-Edo kaimetsu no
     hi,* 24–25. According to Andrew Markus, "One theory . . . maintained that
     Edo was permanently immune from the danger of earthquakes, since a
     vast number of wells . . . provided more than adequate venting for pent-up
     'vapors' in the earth." See "Gesaku Authors and the Ansei Earthquake of
     1855," in Dennis Washburn and Alan Tansman, eds., *Studies in Modern
     Japanese Literature: Essays and Translations in Honor of Edwin McClellan*
     (Ann Arbor: Center for Japanese Studies, University of Michigan, 1997), 55.

110. Subsidence often appeared as "breaking" (*yabure sōrō*), especially in early
     texts. For example, in 1662 the edge of the moat of Osaka Castle "broke" by
     one *shaku.* See "Gen'en jitsuroku," in *DNJS,* vol. 1 (*kō*), 249. After describing
     damage to samurai and commoner structures in the Obama domain of
     Wakasa, one passage explains that "there was 'breakage' to the extent of 3 or 4
     shaku, from which mud flowed" (250).

111. Hatano, Jun, "Edo's Water Supply," in James L. McClain, John W. Merriman, and Ugawa Kaoru, eds., *Edo and Paris: Urban Life and the State in the Early Modern Era* (Ithaca, NY: Cornell University Press, 1994), 247–248.

112. Udagawa Kōsai, *Jishin yobōsetsu* (Edo: Suhara Yaihachi, 1856). Quoted passage, 2; illustration of the rods, 16. See also Hashimoto, *Jishingaku,* 35–36. To view the illustration, see http://archive.wul.waseda.ac.jp/kosho/bunko08/bunko08_c0090/bunko08_c0090_p0019.jpg.

113. Experiments in electricity by Benjamin Franklin and others and the occurrence of several earthquakes in the middle of the eighteenth century prompted scientists such as William Stukeley to posit a connection between seismic activity and electricity in the atmosphere in 1750. A year later, Andrea Bina proposed disequilibria in underground electrical fluid as a cause of earthquakes. Writing in response to the Rimini (Italy) earthquake of 1786, Giuseppe Valadier proposed the construction of giant "para-earthquake" towers that served as lighting rods to channel dangerous electric vapors into the ocean. See Guidoboni and Ebel, *Historical Seismology,* 115, 173–180.

114. Yamazaki Bisei, comp., *Ōjishin rekinenn kō* (1856), in *Edo josei bunko.* Sixth illustration after the table of contents.

115. Hashimoto, *Jishingaku,* 36–38.

116. Explosive theories of earthquakes had a long history in Europe. In *Principia philosophiae* (1644), for example, René Descartes advanced a theory of earthquakes whereby fumes from inside the earth might combine to form combustible mixtures. Soon after the Ansei Edo earthquake, Irish engineer Robert Mallet began important work on seismic waves. Mallet thought that explosive forces of volcanic origin caused earthquakes. See Guidoboni and Ebel, *Historical Seismology,* 167, 170–172, 184.

Chapter 3   Japan according to Earthquakes

1. Kaibara Ekken, *Kadōkun,* Ekken-kai, eds., *Ekken zenshū,* vol. 3 (Ekken zenshū kankōbu, 1911), 452.

2. Print #134 in Miyata Noboru and Takada Mamoru, eds., *Namazue: Shinsai to Nihon bunka* (Ribun shuppan, 1995), 322; see also 10–11. To view this print, see http://gazo.dl.itc.u-tokyo.ac.jp/ishimoto/4/04–017/00002.jpg.

3. Ibid.

4. Ibid.

5. Kitahara Itoko, "Saigai to kawaraban: Sono rekishiteki tenkai," in Kinoshita Naoyuki and Yoshimi Shunya, eds., *Nyūsu no tanjō: Kawaraban to shinbun nishiki-e no jōhōsekai* (Tōkyō daigaku sōgō kenkyū hakubutsukan, 1999), 32–33.

6. Satō Hiroo, *Shinkoku Nihon,* Chikuma shinsho 591 (Chikuma shobō, 2006), 69–73, and Kitai Toshio, *Shinkokuron no keifu* (Hōzōkan, 2006), 50–53.

The *Jōei Formulary* (*Jōei shikimoku*) is more commonly known in Japanese scholarship as *Goseibai shikimoku.*

7. Nagura Tetsuzō, *Fūshigan ishin henkaku: Minshū wa tennō o dō mieta ka* (Kōsō shobō, 2004), 186–187.

8. One catfish print features a temporary brothel named Kannazukiya (House of the month of no deities). See print #116 in Miyata and Takada, *Namazue,* 314–315.

9. Sangawa Akira, *Jishin no Nihonshi: Daichi wa nani o kataru no ka?* (Chūōkōron shinsha, 2007), 169–170.

10. "Shinano bukō, chōshin hikan," in *NRJSS,* vol. 3, 334.

11. "Ansei ni itsubōnen jūgatsu futsuka jishin no koto," in *NJS,* vol. 5, supplement 2, part 1, 535.

12. "Chōsen Taiheiki," in *DNJS,* vol. 1 (*kō*), 207; for other accounts of damage to temples, see 191–193. See also Sangawa, *Jishin no Nihonshi,* 95–96.

13. "Seiu nikki," in *NJS, hoi* (supplement), 175.

14. "Kasubyōshi kan," in *DNJS,* vol. 1 (*kō*), 253–254, and Asai Ryōi, *Kaname'ishi* (1662), in Taniwaki Masachika, Oka Masahiko, and Inoue Kazuhito, eds., trans., *Kanazōshishū* (Shōgakkan, 1999), 35.

15. Asai, *Kaname'ishi* (1662), 40–64, quoted passage, 64. See also Kitahara Itoko, "Saigai to jōhō," in Kitahara Itoko, ed., *Nihon saigaishi* (Yoshikawa kōbunkan, 2006), 236–240.

16. "Yabure mado no ki," in *DNJS,* vol. 2 (*otsu*), 555–556.

17. For modern permutations of the idea of native carpentry practices and architecture as resistant to earthquakes, see Gregory Clancey, *Earthquake Nation: The Cultural Politics of Japanese Seismicity, 1868–1930* (Berkeley: University of California Press, 2006).

18. *AKR,* vol. 1, 4–7 (*jō no yon–jō no shichi*). See also Arakawa Hidetoshi, ed., *Jitsuroku, Ō-Edo kaimetsu no hi: Ansei kenmonroku, Ansei kenmonshi, Ansei fūbunshū* (Kyōikusha, 1982), 29–34.

19. *AKR,* vol. 3, 13–16 (*ge no jūsan–ge no jūroku*). See also Arakawa, *Jitsuroku, Ō-Edo kaimetsu no hi,* 90–94, and Kitahara Itoko, *Jishin no shakaishi: Ansei daijishin to minshū* (Kōdansha, 2000), 188.

20. "Chōryūji monjo," in *NJS, hoi* (supplement), 91.

21. Print #22, Miyata and Takada, *Namazue,* 138, 251–252. See also Wakamizu Suguru, *Edokko kithsitsu to namazue* (Kadokawa gakugei shuppan, 2007), 163–173.

22. *DNJS,* vol. 1 (*kō*), 203–205.

23. Quoted in http://bvd97629.niiblo.jp/e5571.html (accessed November 13, 2010). See also Sangawa, *Jishin no Nihonshi,* 158.

24. See *NJS, hoi* (supplement), 744–745 for other accounts of the earthquake expressed in moral terms. *Tōka nendaiki,* for example, describes the earth

taking revenge on "parents who cast off their children and children who cast off their parents."

25. "Echigo jishin kudoki," at http://blogs.yahoo.co.jp/gojukara11/2937035. html (accessed October 28, 2010). Quoted verses begin thirty-six lines down from the top. See also Hashimoto Manpei, *Jishingaku kotohajime: Kaitakusha Sekiya Seikei no shōgai* (Asahi shinbunsha, 1983), 40–41. In Hashimoto's view, the earthquake was simply a vehicle moralists used to amplify their message.

26. "Echigo jishin kudoki," at http://blogs.yahoo.co.jp/gojukara11/2937035. html (accessed October 28, 2010), and Saitō Masachi, *Goze kudoki jishin no minoue* (publisher unknown, 1829). See also Hashimoto, *Jishingaku,* 40–41, Sankawa, *Jishin no Nihonshi,* 162, and Gerald Groemer, *Bakumatsu no hayari uta: kudokibushi to bushi no shin kenkyū* (Meicho shuppan, 1995), 115–120.

27. Ikeya, for example, claims that the loss of magnetism in the stone at the glasses shop discussed in the previous chapter was "an event known to have occurred two hours before the Ansei Edo Earthquake." See Ikeya Motoji, *Earthquakes and Animals: From Folk Legends to Science* (River Edge, NJ: World Scientific, 2004), vi; see also 5, 12, 13, 15, and 88 for other examples of taking early modern journalistic accounts at face value. Ikeya is hardly alone in this tendency, which I discuss more fully in chapter 6. A true coseismic signal, if such a thing exists, would be a measurable phenomenon that varies in a regular, predictable manner in relation to seismic activity.

28. *AKR,* vol. 1, 1–2 (*jo no ichi, jo no ni*). See also Arakawa, *Jitsuroku, Ō-Edo kaimetsu no hi,* 18.

29. *AKR,* vol. 1, 2 (*jo no san*). See also Arakawa, *Jitsuroku, Ō-Edo kaimetsu no hi,* 22.

30. Kitahara, "Saigai to jōhō," 239.

31. "Chōshin hiroku, jōkan/gekan," in *NRJSS,* vol. 3, 220–222.

32. "Jishin henji go-kyūtōsho," in *NRJSS,* vol. 3, 241–255.

33. "Ansei itsubō jishin kibun," in *NJS,* vol. 5, supplement 2, part 1, 440.

34. "Ansei itsubō jishin kibun," in *NJS,* vol. 5, supplement 2, part 1, 436.

35. For an extensive analysis of the possible reasons for the ban, see Kitahara, *Jishin no shakaishi,* 156–182. Kitahra concludes that *Ansei Chronicle* was not banned because it contained anti-bakufu sentiment, nor because it included some prints that had previously been banned. The main reason was that it was published without approval soon after city authorities had issued repeated prohibitions against such items. In short, its appearance in print at that time (early 1856), not its content, was a de facto provocation.

36. *FN,* 517–518, and *AKS,* vol. 2, 7–9. See also Arakawa *Jitsuroku, Ō-Edo kaimetsu no hi,* 140–142.

37. "Ansei itsubō jishin kibun," in NJS, vol. 5, supplement 2, part 1, 437 and *AKS,* vol. 1, vicinity of p. 15 or 16, near the large illustration of the banks of the

Tenjin River in Honjo in flames. See also Arakawa *Jitsuroku, Ō-Edo kaimetsu no hi*, 121–122, 124 and Noguchi, *Ansei Edo jishin*, 16–17.

38. *AKS*, vol. 3, 16. See also Arakawa *Jitsuroku, Ō-Edo kaimetsu no hi*, 187–188.

39. *AKS*, vol. 2, 6. See also Arakawa *Jitsuroku, Ō-Edo kaimetsu no hi*, 138–139.

40. *AKS*, vol. 3, 21–22 with brief mention in vol. 2, in the text of the illustration of Shin-Yoshiwara located between pp. 2 & 4. See also Arakawa *Jitsuroku, Ō-Edo kaimetsu no hi*, 148–149, 196–197.

41. *AKS*, vol. 3, tale with illustration inserted after p.12 See also Arakawa *Jitsuroku, Ō-Edo kaimetsu no hi*, 182–183.

42. "Ansei itsubō jishin kibun," in *NJS*, vol. 5, supplement 2, part 1, 437 and *AKS*, vol. 1, vicinity of p. 15 or 16, near the large illustration of the banks of the Tenjin River in Honjo in flames. See also Arakawa *Jitsuroku, Ō-Edo kaimetsu no hi*, 121–122, 124 and Noguchi, *Ansei Edo jishin*, 16–17.

43. *AKS*, vol. 1, 16. See also Arakawa *Jitsuroku, Ō-Edo kaimetsu no hi*, 124–125.

44. Kitahara, *Jishin no shakaishi*, 167. For a summary of many of these tales, see pp. 164–167, 206–207.

45. *AKS*, vol. 2, 16. See also Arakawa *Jitsuroku, Ō-Edo kaimetsu no hi*, 143–144.

46. *AKS*, vol. 3, 17. See also Arakawa *Jitsuroku, Ō-Edo kaimetsu no hi*, 189.

47. *AKS*, vol. 3, 17. See also Arakawa *Jitsuroku, Ō-Edo kaimetsu no hi*, 190.

48. *AKS*, vol. 2, 16. See also Arakawa *Jitsuroku, Ō-Edo kaimetsu no hi*, 161–162. See Kitahara, *Jishin no shakaishi*, 167–169 for summaries and a brief discussion of the supernatural tales. Regarding the overall literary nature of the *Kenmonshi*, see Stephan Köhn, "Between Fiction and Non-Fiction: Documentary Literature in the Late Edo Period," in Susanne Formanek and Sepp Linhart, *Written Texts—Visual Texts: Woodblock-Printed Media in Early Modern Japan* (Amsterdam: Hotei Publishing, 2005), 283–310.

49. *Ukiyō no arisama*, quoted in Miki, *Kyōto daijishin*, 74, 105.

50. Miki, *Kyōto daijishin*, 75–78. See pp. 79–86 for the details of other rumors.

51. "Hyōgo kenshi, bekkan" in *NJS*, hoi (supplement), 93.

52. Miki, *Kyōto daijishin*, 87–88.

53. A two-part *kawaraban* print of the Sanjō Earthquake can be found in Nishimaki Kōzaburō, ed. *Kawaraban shinbun: Edo, Meiji sanbyaku jiken*, vol. 1 (Heibonsha, 1978), 61. It claims a wildly exaggerated death toll of 30,000.

54. Kitahara Itoko, *Jishin no shakaishi: Ansei daijishin to minshū* (Kōdansha, 2000), 92–93. For a good example of exaggerated damage reports, see the first sentences of "Jishin kidan miyako manzairaku" in *DNJS*, vol. 1 (kō), 558.

55. Quoted in Kitahara, *Jishin no shakaishi*, 93.

56. Toyo Tokinari, *Honchō jishinki*. Edo josei bunko, vol. 49 (Ōzorasha, 1994), no pagination.

57. Kitahara, *Jishin no shakaishi*, 93.

58. Noguchi, *Ansei Edo jishin*, 214–216.

59. "Tadokoro-shi kiroku," in *DNJS,* vol. 1 (*kō*), 310.

60. For example, see prints #46, #48, and #49 in Miyata and Takada, *Namazue,* 14–15, 223, 267–269. To view print #46, see http://gazo.dl.itc.u-tokyo.ac.jp/ishimoto/2/02-125/00001.jpg. Notice the members of the construction trades in the upper left rushing to protect the giant catfish from attack by Yoshiwara courtesans and employees.

61. *DNJS,* vol. 1 (*kō*), 568.

62. Clancey, *Earthquake Nation,* 131.

63. See J. Charles Schencking, "The Great Kantō Earthquake and the Culture of Catastrophe and Reconstruction in 1920s Japan," *Journal of Japanese Studies* 34, no. 2 (summer 2008): 295–331 (quote on 303).

64. "Tadokoro-shi kiroku," in *DNJS,* vol. 1 (*kō*), 310. A *koku* was roughly 180 liters in volume or 150 kilograms in weight.

65. "Tanabe mandaiki," in *NRJSS,* vol. 2, 90–92.

66. Kitahara Itoko, "Kinsei ni okeru saigaikyūsai to fukkō," in Kitahara, *Nihon saigaishi,* 196–197.

67. "Gokiroku kenbyōki," in *NRJSS,* vol. 2, 171.

68. "Senfukuji shoji kenmon zakki," in *NRJSS,* vol. 2, 175.

69. For details, including extensive tabular data, see *NRJSS,* vol. 2, 179–190. For additional documentation of local relief efforts, see 191–199.

70. Harada Kazuhiko, "Matsushiro-han," in Chūō bōsai kaigi, *1847 Zenkōji jishin,* 110.

71. For a hand-drawn map illustrating the collapsed mountain and affected villages, see *NRJSS,* vol. 2, 239–240. For a better version of this same map and another similar to it, see *NRJSS,* vol. 3, 354–357. For another map, see http://www.um.u-tokyo.ac.jp/publish_db/1999news/02/images/030_01.jpg.

72. "Kenshūroku," in *DNJS,* vol. 2 (*otsu*), 105–107, 109–111; Harada Kazuhiko, "Matsushiro-han," in Chūō bōsai kaigi, *1847 Zenkōji jishin,* 110–111; Itō Kazuaki, *Jishin to funka no Nihonshi* (Iwanami shoten, 2002), 138–139; Shinano mainichi shinbunsha kaihstsukyoku shuppanbu, *Zenkōji daijishin,* 138–150; and Kitō Yasuyuki and Nagase Satoshi, "Saigai to kyūsai: Machi to mura," in Akahane and Kitahara, *Zenkōji jishin,* 75–94.

73. "Kenshūroku," in *DNJS,* vol. 2 (*otsu*), 107, 109–111; Itō, *Jishin to funka no Nihonshi,* 139–141; and Shinano mainichi shinbunsha kaihstsukyoku shuppanbu, *Zenkōji daijishin,* 150–180.

74. Harada, "Matsushiro-han," 111. For a comprehensive study of emergency response, relief, and rebuilding efforts, see 71–121. See also Shinano mainichi shinbunsha kaihstsukyoku shuppanbu, *Zenkōji daijishin,* 180–204. For a sample of relevant documents concerning specific relief efforts and proposals, see "Kamahara Dōzan jishin kiji," 51–52, 83–85, 90–91; "Nogisono zakki,"

139; "Daijishin kōzui saigai kiroku," 212–232; and "Sinkōkan," 317–318, in *DNJS*, vol. 2 (*otsu*).

75. Itō, *Jishin to funka no Nihonshi*, 186, and Kitahara, "Saigai to jōhō," 234. The process of creating new fields and preserving them also required active dredging and infrastructure work by the Obama domain. For details, see Higashi Sachiyo, "Uramikawa kussaku jigyō to shinden kaihatsu," in Chūō bōsai kaigi, *1662 Kanbun Ōmi-Wakasa jishin hōkokusho* (Mizuho jōhōsōken kabushikigaisha, 2005), 115–119.

76. Higashi Sachiyo, "Kanbun jishin ga motarashita mono," in Chūō bōsai kaigi, *1662 Kanbun Ōmi-Wakasa jishin hōkokusho*, 119–121; tabular data, 120.

77. "Mikata goko shūhen no shinden kaihatsu," in *NJS*, hoi (supplement), 180.

78. "Higata Kanzeon engi," in *NRJSS*, vol. 2, 61.

79. Sangawa, *Jishin no Nihonshi*, 137. The geologically similar 1923 Great Kantō Earthquake caused approximately two meters of uplift.

80. Itō, *Jishin to funka no Nihonshi*, 86–88. Itō states that maximum uplift of the Bōsō Peninsula was 5.5 meters.

81. "Tateyama-wan engan no okeru Genroku jishin higai to higata riyō nitsuite," in *NRJSS*, vol. 3, 122–125. See these pages for additional examples and maps.

82. For details, see "Tateyama-wan engan ni okeru Genroku jishin higai to higata riyō nitsuite," in *NRJSS*, vol. 3, 125–128.

83. Sangawa, *Jishin no Nihonshi*, 149–154, and Itō, *Jishin to funka no Nihonshi*, 121–126.

84. Nishimaki, *Kawaraban shinbun*, 125, and Inagaki, *Edo no taihen*, 50–51.

85. Sangawa, *Jishin no Nihonshi*, 180–181, and Ishibashi, *Daishinran no jidai*, 25–26.

86. Ishibashi, *Daishinran no jidai*, 27.

87. "Kiki kōki," in *DNJS*, vol. 1 (*kō*), 284.

88. "Kasshi yawa," in *DNJS*, vol. 1 (*kō*), 523.

89. Susanna M. Hoffman, "The Monster and the Mother: The Symbolism of Disaster," in Susanna M. Hoffman and Anthony Oliver-Smith, eds., *Catastrophe and Culture: The Anthropology of Disaster* (Santa Fe, NM: School of American Research Press, 2002), 131.

90. "Chōshin hiroku, gekan," in *NRJSS*, vol. 3, 237–241.

91. Okumiya Masaaki, *Kokuryōki* (no date or publisher), nineteenth page face, or "Kokuryōki," in *DNJS*, vol. 1 (*kō*), 321. Regarding the Hakuhō Earthquake, see Itō, *Jishin to funka no Nihonshi*, 8–10. A second entry in Kokuryōki also describes the Hakuhō earthquake and the massive inundation of fields, which "since then have been the ocean." The passage explains that knowledge of the Hakuhō earthquake has been passed down in Tosa as local legend but without definite proof. In any case, the present earthquake is "unprecedented

since ancient times." "Kokuryōki" (the second "Kokuryōki" entry), in *DNJS*, vol. 1 (*kō*), 330. For a diagram showing land loss in Tosa before and after the Hakuhō earthquake, see *NJS, hoi* (supplement), 2–3.

92. Print #73, in Miyata and Takada, *Namazue,* 288–291.

93. See Yan Y. Kagan, David G. Jackson, and Robert J. Geller, "Characteristic Earthquake Model, 1884–2011, R.I.P.," *Seismological Research Letters* 83 (November/December 2012): 951–953, and Robert Geller (Robaato Geraa), *Nihonjin wa shiranai "jishin yochi" no shōtai* (Futabasha, 2012), 160–161.

94. Hoffman, "The Monster and the Mother," 133.

95. "Gen'en jitsuroku," in *DNJS*, vol. 1 (*kō*), 250–251. The reign name date for the first Chinese earthquake, "Xiangxing 27," is impossible because this reign lasted only one year, from 1278 to 1279. Asai Ryōi also mentions the same Chinese earthquake that killed seven thousand. See Asai, *Kaname'ishi,* 69.

96. "Kōretsu hikki" in *DNJS*, vol. 1 (*kō*), 329.

97. "Zoku jishin zassan," in *DNJS*, vol. 2 (*otsu*), 428.

98. "Koizumi Sōken nichiroku," in *NRJSS*, vol. 2, 233.

99. *DNJS,* vol. 1 (*kō*), 568.

100. "Kanrosō," in *DNJS*, vol. 1 (*kō*), 302.

101. "Jishin kidan miyako manzairaku," in *DNJS*, vol. 1 (*kō*), 558, 560. There was a long-standing folk custom of singing this verse to ward off earthquakes. See Koga rekishi hakubutsukan, eds., *Tenpenchii to seikimatsu: Nihonjin no saigaikan* (Ibaraki-ken, Koga-shi, Japan: Koga rekishi hakubutsukan, 1999), 35.

102. "Shinano bukō, Chōshin hikan," in *NRJSS*, vol. 3, 313.

103. "Eikan zasshi," in *DNJS*, vol. 2 (*otsu*), 248–249.

104. Satō, *Shinkoku Nihon,* 90–93, 130–158.

105. "Ōsaka jishinki," in *DNJS,* vol. 2 (*otsu*), 334.

106. "Zoku jishin zassan," in *DNJS*, vol. 2 (*otsu*), 419.

107. "Zoku jishin zassan," in *DNJS*, vol. 2 (*otsu*), 455.

108. *NRJSS,* vol. 2, 434.

Chapter 4   The Ansei Edo Earthquake

1. "Ōjishin ōkaze kenmonroku," in *NJS*, vol. 5, supplement 2, part 1, 533.

2. "Nai no nochimigusa," in *DNJS*, vol. 2 (*otsu*), 579–580.

3. "Jifūroku," in *DNJS*, vol. 2 (*otsu*), 538. See also "Ansei itsubō jishin kibun," in *NJS*, vol. 5, supplement 2, part 1, 446.

4. Chūō bōsai kaigi, eds., *1855 Ansei Edo jishin hōkokusho* (Fuji sōgō kenkyūsho, 2004), 77.

5. *AKR,* vol. 3, 10 (*ge no jū*). See also Arakawa Hidetoshi, ed., *Jitsuroku, Ō-Edo kaimetsu no hi: Ansei kenmonroku, Ansei kenmonshi, Ansei fūbunshū* (Kyōikusha, 1982), 85–86.

6. According to Edward A. Keller and Nicholas Pinter, "The potential for amplification of surface waves to cause damage was again demonstrated with tragic results in the 1989 M_w 7.2 Loma Prieta (San Francisco) earthquake, when the upper tier of the Nimitz Freeway in Oakland, California, collapsed, killing forty-one people (fig. 1.16). Collapse of the tiered freeway occurred on a section of roadway constructed on bay fill and mud. Where the freeway was constructed on older, stronger alluvium, less shaking occurred and the structure survived. Extensive damage was also recorded in the Marina District of San Francisco (fig. 1.17), primarily in areas constructed on bay fill and mud, including debris dumped into the bay during the cleanup following the 1906 earthquake." See Keller and Pinter, *Active Tectonics: Earthquakes, Uplift, and Landscape*, 2nd ed. (Upper Saddle River, NJ: Prentice Hall, 2002), 21–23 (figures on 24).

7. Susan Elizabeth Hough, *Earthshaking Science: What We Know (and Don't Know) about Earthquakes* (Princeton, NJ: Princeton University Press, 2002), 80–83.

8. Bruce A. Bolt, *Earthquakes* (New York: W. H. Freeman and Company, 1993, fifth printing, 1997), 266.

9. Bolt, *Earthquakes,* 132.

10. Kitahara Itoko, ed., *Nihon Saigaishi* (Yoshikawa kōbunkan, 2006), 254.

11. Andrew L. Markus, "Gesaku Authors and the Ansei Earthquake of 1855," in Dennis Washburn and Alan Tansman, eds., *Studies in Modern Japanese Literature* (Ann Arbor: Center for Japanese Studies, University of Michigan, 1997), 55.

12. Gregory Clancey, *Earthquake Nation: The Cultural Politics of Japanese Seismicity, 1868–1930* (Berkeley: University of California Press, 2006), 123, 220.

13. Ibid., 123.

14. Most major studies of the Ansei Edo earthquake from the 1990s onward discuss military casualties in some detail. The most extensive treatment, including discussion of issues in interpreting different types of data, is Chūō bōsai kaigi, *1855 Ansei Edo jishin hōkokusho* (Fuji sōgō kenkyūsho, 2004), especially Kitahara Itoko, "Saigai to Shakaizō," 43–59. See also the detailed tables and charts in this volume, 95–122. Also valuable is Noguchi Takehiko, *Ansei Edo jishin: Saigai to seiji kenryōku* (Chikuma shobō, 1997), 72–73, 108–110.

15. Clancey, *Earthquake Nation,* 55.

16. Nakamura Misao, "Ansei Edo jishin," in Chūō bōsai kaigi, *1855 Ansei Edo jishin hōkokusho,* 2–5.

17. Kitahara, "Saigai to Shakaizō," 51.

18. Noguchi, *Ansei Edo jishin,* 71, 73, 75 (map), 91–92. For an extended discussion of the interconnections between topography and political power

with respect to the destruction of prime bakufu real estate, see 71–128. Subsequent studies have reinforced many of Noguchi's main points. See Nakamura Misao, Matsuura Ritsuko, Kayano Ichirō, Karakama Ikuo, and Nishiyama Akihito, "Ansei Edo jishin (1855/11/11) no Edo shichū no higai," *Rekishi jishin* 18 (2002): 77–96, and Nakamura, Kayano, and Matsuura, "Ansei Edo Jishin no shutoken de no higai," *Rekishi jishin* 19 (2003): 32–37.

19. Saitō Gesshin, *Ansei itsubō bukō chidō no ki,* in Edo sōsho kankō kai, eds., *Edo sōsho 9* (Edo sōsho kankōkai, 1917), 2.

20. For a discussion of these geographical details, see Noguchi, *Ansei Edo jishin,* 73–81, 97–98. Especially helpful are the maps on 75, 79, and 99. Also helpful for visualizing the differences between pre-1600 Edo and post-1600 Edo are the two maps on the inside cover of Akira Naito, *Edo, the City That Became Tokyo: An Illustrated History,* Kazuo Hozumi, illus., H. Mack Horton, trans. (Kodansha, 2003).

21. See Naito, *Edo,* 36–37, for more details. See also Tsuji Yoshinobu, *Sennen shinsai: Kurikaesu jishin to tsunami no rekishi ni manabu* (Daiyamondo sha, 2011), 88–90, 182, for a useful discussion and map of daimyō mansions and Hibiya Cove.

22. Noguchi, *Ansei Edo jishin,* 98.

23. "Yabure mado no ki," in *DNJS,* vol. 2 (*otsu*), 559–560.

24. Sakuma Chōkei, "Ansei daijishin jikken dan," in *NJS,* vol. 5, supplement 2, part 1, 470.

25. Noguchi, *Ansei Edo jishin,* 95–96.

26. "Yabure mado no ki," in *DNJS,* vol. 2 (*otsu*), 561.

27. "Jifūroku," in *DNJS,* vol. 2 (*otsu*), 538.

28. "Nai no nochimigusa," in *DNJS,* vol. 2 (*otsu*), 573–574.

29. "Nai no nochimigusa," in *DNJS,* vol. 2 (*otsu*), 574. See also Noguchi, *Ansei Edo jishin,* 72–73, for discussion of another source reaching precisely the same conclusions, and Kitahara Itoko, *Jishin no shakaishi: Ansei daijishin to minshū* (Kōdansha, 2000), 109–111. See also Tsuji, *Sennen shinsai,* 88–90.

30. "Nai no nochimigusa," in *DNJS,* vol. 2 (*otsu*), 572.

31. For discussions of secondary and tertiary factors such as population density, construction methods, access to open spaces, and so forth, see Kitahara, "Saigai no shakaizō," 53, 60–64, and Kitahara Itoko, *Jishin no shakaishi,* 50–79.

32. "Ansei daijishin jikken dan," in *NJS,* vol. 5, supplement 2, part 1, 473–474. See also Kitahara, "Saigai no shakaizō," 64.

33. Saitō, *Ansei itsubō bukō chidō no ki,* 4. See also Noguchi, *Ansei Edo jishin,* 33–34, 148–149.

34. For the initial survey order and preliminary data, including property damage, see *NJS,* vol. 5, supplement 2, part 1, 52, 53–56. See also 59–60 regarding follow-up and 90–95 for decrees from the Machigaisho and preliminary data.

35. Kitahara, *Jishin no shakaishi,* 45–52, 71, 74–79, 114–115 (regarding survey results appearing in the press). The data for casualties appears in tabular form on 46, and an image of the popular press report of casualty figures is on 113. See also Kitahara, "Saigai no shakaizō," 59–60, with the data in tabular form in Chūō bōsai kaigi, *1855 Ansei Edo jishin hōkokusho,* 97. This case was probably not the first instance of bakufu officials providing government information to popular publishers. In the context of discussing commercial publication since the middle seventeenth century of an encyclopedic genre known as Military Mirrors (*bukan*), Mary Elizabeth Berry concludes that "the Tokugawa administration itself was the major supplier of Mirror material." See Berry, *Japan in Print: Information and Nation in the Early Modern Period* (Berkeley: University of California Press, 2006), 110.

36. Saitō, *Ansei itsubō bukō chidō no ki,* 23–24, and "Fujiokaya nikki (*ge*)," in *NJS,* vol. 5, supplement 2, part 1, 410–412. Kitahara bases her analysis of casualties on the figures of 4,293 deaths and 2,759 injuries. See Kitahara, *Jishin no shakaishi,* 46, and Chūō bōsai kaigi, *1855 Ansei Edo jishin hōkokusho,* 97, for a presentation of the data in tabular form.

37. "Nai no nochimigusa," in *DNJS,* vol. 2 (*otsu*), 574–579.

38. "Yabure mado no ki," in *DNJS,* vol. 2 (*otsu*), 556–557, 558.

39. "Jifūroku," in *DNJS,* vol. 2 (*otsu*), 539–542.

40. The 540,000 figure comes from Kitahara, "Saigai no shakaizō," 60.

41. "Kainai jishinroku," in *NJS,* vol. 5, supplement 2, part 1, 503.

42. "Ansei itsubō Edo daijishin hikki," in *NJS,* vol. 5, supplement 2, part 1, 516.

43. "Yabure mado no ki," in *DNJS,* vol. 2 (*otsu*), 558.

44. Anonymous, *Jishin narabini shukka saikenki* (publisher unknown, 1855), image spanning 7–8. To view the image, see http://archive.wul.waseda.ac.jp/kosho/wo01/wo01_02952/wo01_02952_p0009.jpg.

45. *AKS,* vol. 3, illustrations by Utagawa Kuniyoshi et al., author(s) and publisher unknown, 1856, twenty-fourth and twenty-fifth page faces. To view the image, see http://archive.wul.waseda.ac.jp/kosho/wo01/wo01_04209/wo01_04209_0003/wo01_04209_0003_p0014.jpg.

46. "Ansei itsubō Edo daijishin hikki," in *NJS,* vol. 5, supplement 2, part 1, 516.

47. "Jifūroku," in *DNJS,* vol. 2 (*otsu*), 539.

48. "Ansei itsubō jishin kibun," in *NJS,* vol. 5, supplement 2, part 1, 433.

49. Kitahara, *Jishin no shakaishi,* 111–112.

50. For a map of the originally planned eleven batteries, see Nagura Tetsuzō, *Fūshigan ishin henkaku: Minshū wa tennō o dō mieta ka* (Kōsō shobō, 2004), 168.

51. Noguchi, *Ansei Edo jishin,* 61, 64.

52. Nagura, *Fūshigan ishin no henkaku,* 171–173. The popular verse was based on *Hyanunin isshu* #15.

53. Noguchi, *Ansei Edo jishin,* 103–104.
54. One example is "Ansei itsubō Edo daijishin hikki," in *NJS,* vol. 5, supplement 2, part 1, 516. The story of the fire in the magazine is preceded by a disclaimer in small letters: "This is probably a groundless tale." See also "Ansei itsubō jishin kibun," in *NJS,* vol. 5, supplement 2, part 1, 438, 451. That account's author, Miyazaki Narumi, includes the tale of the exploding magazine in volume 1 of his account and corrects it in volume 2.
55. For example, see *Edo Ōjishin matsudai hanashi no tane,* 2. To view this image, see http://archive.wul.waseda.ac.jp/kosho/wo01/wo01_03639/wo01_03639_p0003.jpg.
56. Ansei itsubō jishin kibun," in *NJS,* vol. 5, supplement 2, part 1, 451. Platforms 1, 2, and 3 "were shaken down into the ocean."
57. Noguchi, *Ansei Edo jishin,* 103–107.
58. "Ansei itsubō jishin kibun," in *NJS,* vol. 5, supplement 2, part 1, 438.
59. Markus, "Gesaku Authors," 57.
60. "Nai no nochimigusa," in *DNJS,* vol. 2 (*otsu*), 573.
61. "Jifūroku," in *DNJS,* vol. 2 (*otsu*), 538.
62. Saitō, *Ansei itsubō bukō chidō no ki,* 9. See Markus, "Gesaku Authors," 57–58, for a lengthy, vivid description by Saitō Gesshin from a different work.
63. "Fujiokaya nikki (*ge*)," in *NJS,* vol. 5, supplement 2, part 1, 386.
64. "Ansei itsubō jishin kibun," in *NJS,* vol. 5, supplement 2, part 1, 443. See also Noguchi, *Ansei Edo jishin,* 141.
65. *NJS,* vol. 5, supplement 2, part 1, 73–74.
66. Cecilia Segawa Seigle, *Yoshiwara: The Glittering World of the Japanese Courtesan* (Honolulu: University of Hawai'i Press, 1993), 207.
67. For a list of temporary brothel locations, see print #150, Miyata and Takada, *Namazue,* 332–333; "Yabure mado no ki," in *DNJS,* vol. 2 (*otsu*), 565–566; and *NJS,* vol. 5, supplement 2, part 1, 73.
68. The basic cost at these brothels was 1 *bu* (one-fourth of a *ryō*). See Kitani Makoto, *Namazue shinkō: Saigai no kosumorojii* (Tsuchiura-shi: Tsukuba shorin, 1984), 57. For comparison, the basic cost of a night with a *zashikimochi* (one rank blow an *oiran*) at Shin-Yoshiwara was 1 or 2 *bu,* and a night with a *heyamochi* (one rank lower) ranged from 0.5 *bu* to 1.0 *bu.* See Cecilia Segawa Seigle, *Yoshiwara: The Glittering World of the Japanese Courtesan* (Honolulu: University of Hawai'i Press, 1993), 231.
69. "Jishin kidan roku," in *NJS,* vol. 5, supplement 2, part 1, 508. This tale is reminiscent of an account by Asai Ryōi in 1662 of a collapsing storehouse killing all the women of a wealthy household. See Asai Ryōi, *Kaname'ishi* (1662), in Taniwaki Masachika, Oka Masahiko, and Inoue Kazuhito, eds., trans., *Kanazōshishū* (Shōgakkan, 1999), 21–22.
70. "Nai no nochimigusa," in *DNJS,* vol. 2 (*otsu*), 570.

71. Noguchi, *Ansei Edo jishin*, 118–120.

72. "Ansei itsubō jishin kibun," in *NJS*, vol. 5, supplement 2, part 1, 432–433. See also 454, 455. Noguchi points out that according to official allowances, the wages of roof thatchers rose 50 percent, while roof tile installers' wages rose only 9 percent because tiled roofs lost their popularity. *Ansei Edo jishin*, 202. Actual wage rates were higher, but the official figures indicate the approximate relative differences.

73. See, for example, "Ansei itsubō jishin kibun," in *NJS*, vol. 5, supplement 2, part 1, 431, 441–442, 454, and 455–456.

74. Noguchi, *Ansei Edo jishin*, 217.

75. Print #23 in Miyata and Takada, *Namazue*, 252. See also Wakamizu Suguru, *Edokko kishitsu to namazue* (Kadokawa gakugei shuppan, 2007), 71–73. See figure 7 in the present volume.

76. "Yabure mado no ki," in *DNJS*, vol. 2 (*otsu*), 552–553.

77. "Jifūroku," in *DNJS*, vol. 2 (*otsu*), 539.

78. "Ansei itsubō jishin kibun," in *NJS*, vol. 5, supplement 2, part 1, 432, 454.

79. Print #62 in Miyata and Takada, *Namazue*, 18–19, 278–280. See also Wakamizu Suguru, *Namazu wa odoru: Edo no namazue omoshiro bunseki* (Bungeisha, 2003), 62–65.

80. See, for example, prints #110 and #111 in Miyata and Takada, *Namazue*, 138, 311–312. See also Markus, "Gesaku Authors," 56–57, and Kitahara, *Jishin no shakaishi*, 241–245.

81. *Nangitori* (Hard-to-Figure-Out Bird) is a good example of a print critical of the newly rich. Five tradesmen are sitting around in an expensive restaurant. A large catfish is going to be their feast during a night of drinking and revelry. A giant bird, however, swoops down and snatches the catfish away from them. The bird is "hard to figure out" because it consists entirely of tools and objects from occupations adversely effected by the earthquake. Its tail feathers are oars for small boats, and its wings are books, dry goods, abacuses, and tall *geta* shoes. Its neck and crown are the hairpins of elite courtesans and tea ceremony whisks. In other words, the bird represents such professions as tea ceremony teachers, courtesans, and small boat operators. It also includes booksellers, pawnshops, and clothing stores. Print #108 in Miyata and Takada, *Namazue*, 225, 310–311. To view this print, see http://gazo.dl.itc.u-tokyo.ac.jp/ishimoto/2/02-058/00001.jpg. See also Tomisawa Tatsuzō, "Nishikie no nyūsu sei: Namazue, hashikae, Bōshin sensō-ki no fūshiga o megutte," in Kinoshita Naoyuki and Yoshimi Shunya, eds., *Nyūsu no tanjō: Kawaraban to shinbun nishikie no jōhō sekai* (Tōkyō daigaku sōgō kenkyū hakubutsukan, 1999), 195.

82. Print #88 in Miyata and Takada, *Namazue*, 224, 298–299. For a discussion of prints featuring this motif, see Wakamizu, *Edokko*, 60–65. To view this print, see http://gazo.dl.itc.u-tokyo.ac.jp/ishimoto/2/02-100/00001.jpg.

83. Print #99 in Miyata and Takada, *Namazue*, 224, 303–304. See also Abe Yoshinari, "Jishin to hitobito no sōzōryoku," in Chūō bōsai kaigi, *1855 Ansei Edo jishin hōkokusho,* 141–142. To view this print, see http://gazo.dl.itc .u-tokyo.ac.jp/ishimoto/2/02–035/00001.jpg.

84. Print #94 in Miyata and Takada, *Namazue,* 301.

85. Print #85 in Miyata and Takada, *Namazue,* 297. See also Abe Yasunari, "Jishin to hitobito no sōzōryoku" in Chūō bōsai kaigi, *Ansei Edo jishin,* 139–143. To view this print, see http://dl.ndl.go.jp/info:ndljp/pid/1302029.

86. Print #130 in Miyata and Takada, *Namazue,* 13, 320–322. See figure 9 in the present volume.

87. Noguchi, *Ansei Edo jishin, hyō* 8, 203.

88. "Ansei itsubō jishin kibun," in *NJS,* vol. 5, supplement 2, part 1, 456; Noguchi, *Ansei Edo jishin,* 202–204; Kitahara, *Jishin no shakaishi,* 245–246; and Kitahara, "Saigai no shakaizō," 66.

89. "Ansei daijishin jikkendan," in *NJS,* vol. 5, supplement 2, part 1, 456.

90. See Kitahara, *Jishin no shakaishi,* 245–246, for a discussion of this matter.

91. Print #129 in Miyata and Takada, *Namazue,* 222, 320. To view this image, see http://www.um.u-tokyo.ac.jp/publish _db/1999news/04/403/images/194.jpg.

92. Print #69 in Miyata and Takada, *Namazue,* 285–286. To view this print, see http://gazo.dl.itc.u-tokyo.ac.jp/ishimoto/2/02–047/00001.jpg.

93. Saitō, *Ansei itsubō bukō chidō no ki,* 29. See also "Jifūroku" *DNJS,* vol. 2 (*otsu*), 539.

94. This round of temple services was the second major religious intervention by the bakufu. On the seventh day of the tenth month, it paid thirteen shrines throughout the country to conduct earthquake prayer rites. See Noguchi, *Ansei Edo jishin,* 194.

95. *NJS,* vol. 5, supplement 2, part 1, 69–70.

96. For brief letters by retainers to the *ōmetsuke,* reporting on their actions the night of the earthquake, see "Bakufu satasho," in *DNJS,* vol. 2 (*otsu*), 527–537. Some also include informal mention of losses incurred by the retainers. See also Kitahara, "Saigai no shakaizō," 43–47, and Noguchi, *Ansei Edo jishin,* 83–84.

97. *NJS,* vol. 5, supplement 2, part 1, 3.

98. Ibid.

99. Noguchi, *Ansei Edo jishin,* 181–184, and Kitahara, "Saigai no shakaizō," 47–50.

100. Noguchi, *Ansei Edo jishin,* 187–189, and Kitahara, "Saigai no shakaizō," 46.

101. For specific letters authorizing the loans and grants, including the dates and amounts of the loans, see Tōkyō shiyakusho, ed., *Tōkyō shishi kō, kyūsaihen* (Rinsen shoten, 1975), 457–467 and 160–163, for official entries regarding these matters. See also Kitahara, "Saigai no shakaizō," 50–51, 67–68, and *hyō* 2-8, 2-9, and 2-10; Chūō bōsō kaigi, *1855 Ansei Edo jishin hōkokusho,* 99;

and Noguchi, *Ansei Edo jishin,* 184–188, including *hyō* 6, 7, and 8. See also "Jifūroku," in *DNJS,* vol. 2 (*otsu*), 539.

102. Noguchi, *Ansei Edo jishin,* 187–188.

103. Information based on the recollections of Sakuma Chōkei. See "Ansei daijishin jikkendan," in *NJS,* vol. 5, supplement 2, part 1, 470. See also Kitahara, *Jishin no shakaishi,* 252, and Noguchi, *Ansei Edo Jishin,* 149–150.

104. *NJS,* vol. 5, supplement 2, part 1, 56–57.

105. Ibid., 61–62. See also Kitahara, "Saigai no shakaizō," 74–75, and Kitahara, *Jishin no shakaishi,* 259–260, 266–268.

106. For details regarding medical treatment, see "Ansei daijishin jikkendan," in *NJS,* vol. 5, supplement 2, part 1, 472–473.

107. *NJS,* vol. 5, supplement 2, part 1, 79.

108. For a discussion of these practical difficulties, see "Jifūroku," in *DNJS,* vol. 2 (*otsu*), 539.

109. *NJS,* vol. 5, supplement 2, part 1, 90.

110. "Ansei daijishin jikkendan," in *NJS,* vol. 5, supplement 2, part 1, 470–471; Noguchi, *Ansei Edo jishin,* 155–160; and Kitahara, *Jishin no shakaishi,* 254–263. For tables showing the locations and number of residents in temporary housing, see Kitahara, 254 and *hyō* 2-13, Chūō bōsō kaigi, *1855 Ansei Edo jishin hōkokusho,* 101. Nearly every substantial account of the earthquake and many popular prints list the locations of temporary housing. To cite but one example, see "Yabure mado no ki," in *DNJS,* vol. 2 (*otsu*), 554.

111. *NJS,* vol. 5, supplement 2, part 1, 97–98.

112. "Ansei daijishin jikkendan," in *NJS,* vol. 5, supplement 2, part 1, 473. There is an excellent visual image of a similar scene toward the end of a long scroll of earthquake-related materials from 1855: *Edo ōjishin kiji.* To view the image, see http://archive.wul.waseda.ac.jp/kosho/bunko10/bunko10_08871/bunko10_08871_p0018.jpg. Two samurai officials sit with a banner that says "*go-yō*" (on duty) as other officials distribute food aid (directly above it is an announcement about temporary housing). *AKR* includes an image of a samurai official distributing food aid in the form of cooked rice (in front of a damaged restaurant). *AKR,* vol. 1, 4 (*jō e yon*). To view this image, see http://archive.wul.waseda.ac.jp/kosho/wo01/wo01_03628/wo01_03628_0001/wo01_03628_0001_p0023.jpg. See also Noguchi, *Ansei Edo jishin,* 159–160.

113. *NJS,* vol. 5, supplement 2, part 1, 97.

114. Ibid., 97–98.

115. For documents connected with food relief, see "Shimai, hidari no gotoshi," in Tōkyō shiyakusho, ed., *Tōkyōshi shikō, kyūsaihen 4* (Rinsen shoten, 1975, originally published 1922), 474–487, and *NJS,* vol. 5, supplement 2, part 1, 95–105, 180–182. See also Kitahara, *Jishin no shakaishi,* 260–265, and Kitahara, "Saigia no shakaizō," 75–76.

116. The broader context was the shogunate's Edo-first policy. See Anne Walthall, "Edo Riots," in James L. McClain, John W. Merriman, and Ugawa Kaoru, eds., *Edo and Paris: Urban Life and the State in the Early Modern Era* (Ithaca, NY: Cornell University Press, 1994), 419.

117. *NJS,* vol. 5, supplement 2, part 1, 52–53. See also 63–64 regarding lumber supplies.

118. Kitahara, "Saigia no shakaizō," 76–78, and Kitahara, *Jishin no shakaishi,* 246–248.

119. For the various decrees concerning wages and prices, see *NJS,* vol. 5, supplement 2, part 1, 51, 58–59, 62, 64–72, 79–80, and 90.

120. *NJS,* vol. 5, supplement 2, part 1, 79–80. A similar decree was issued on the same day (supplemented on the next day) warning plasterers about excessive wages. See 80–81.

121. Regarding righteous granaries, see Mark J. Ravina, "Confucian Banking: The Community Granary (*Shasō*) in Rhetoric and Practice," in Bettina Gramlich-Oka and Gregory Smits, eds., *Economic Thought in Early Modern Japan* (Leiden: Brill, 2010), 179–204.

122. *NJS,* vol. 5, supplement 2, part 1, 107–121.

123. Ibid., 74.

124. Ibid., 68–69. See also Kitahara, "Saigai no shakaishi," 78.

125. For a warning against such fraud, issued on the fifth day of the thirteenth month, see *NJS,* vol. 5, supplement 2, part 1, 74.

126. Kitahara, *Jishin no shakaishi,* 167.

127. *AKS,* vol. 3, 5–6. See also Arakawa, *Jitsuroku, Ō-Edo kaimetsu no hi,* 172.

128. Kitahara, *Jishin no shakaishi,* 288–290. For several examples of *bushi* contributions, see "Ansei zakki," in *NJS,* vol. 5, supplement 2, part 1, 183–184.

129. Kitahara, *Jishin no shakaishi,* 126, 286–287.

130. For examples of dozens of temples actively involved in aid, see "Ansei zakki," in *NJS,* vol. 5, supplement 2, part 1, 184–189, 193–196, 198–200, 210–214, 217–222, and 224–231.

131. Kitahara, *Jishin no shakaishi,* 288, and Noguchi, *Ansei Edo jishin,* 194.

132. For decrees and the text of formal notices of award, see *NJS,* vol. 5, supplement 2, part 1, 76–78.

133. *NJS,* vol. 5, supplement 2, part 1, 125. For documents recognizing hundreds of examples of relief provided by a wide range of townspeople, see *NJS,* vol. 5, supplement 2, part 1, 122–157.

134. Kitahara, "Saigai no shakaizō," 79.

135. *AKS,* vol. 1, 6, 8–11, 14, vol. 2, 10–13, 5 and 6 page faces from end (pagination barely discernible), and vol. 3, 7–9, 11. See also Arakawa, *Jitsuroku, Ō-Edo kaimetsu no hi,* 107–108, 109, 113, 114–115, 117, 151, 155–155, 168, 169, 173, and 177.

136. Noguchi, *Ansei Edo jishin,* 174, 176–179. Kitahara points out that popular newspapers issued *segyō* (charity) editions to honor donors. Such publicity, of course, also had the effect of pressuring others to contribute. *Jishin no shakaishi,* 272.

137. Kitahara, *Jishin no shakaishi,* 282–283, 304–305.

138. Ibid., 274–275.

139. Kitahara, "Saigai no shakaizō," 79–80. For data on cash donations in tabular form, see *hyō* 2-13, Chūō bōsō kaigi, *1855 Ansei Edo jishin hōkokusho,* 101.

140. Kitagara, *Jishin no shakaishi,* 272–274, and Noguchi, *Ansei Edo jishin,* 179–181.

141. Kitahara, *Jishin no shakaishi,* 307–327. For the definition of *girei,* see 307, and for the links between *segyō* and *okage-mairi,* see 326–328.

142. "Ansei itsubō jishin kibun," in *NJS,* vol. 5, supplement 2, part 1, 441–442.

143. Kitahara, "Saigai no shakaizō," 81–83, 86.

144. For documents detailing post-earthquake repairs, see *NJS,* vol. 5, supplement 2, part 1, 166–176, and Kitahara, "Saigai no shakaizō," 84–88. For details on what was repaired when, see *zu* 2-17 and 2-18 in Chūō bōsō kaigi, *1855 Ansei Edo jishin hōkokusho,* 118, 199.

145. Kitahara, "Saigai no shakaizō," 86, 88–92.

Chapter 5   Meanings

1. Print #175 in Miyata Noboru and Takada Mamoru, eds., *Namazue: shinsai to Nihon bunka* (Ribun shuppan, 1995), 347–348. See also Suehiro Sachiyo, "Ōtsue no hyōtan-namazu," in Miyata and Takada, *Namazue,* 182–185. To view this print, see http://metro2.tokyo.opac.jp/tml/tpic/imagedata/toritsu/ukiyoe/0C/0277-C014.jpg.

2. "Kainai jishinroku," in *NJS,* vol. 5, supplement 2, part 1, 503.

3. Print #195 in Miyata and Takada, *Namazue,* 222, 356–357. To view this print, see http://gazo.dl.itc.u-tokyo.ac.jp/ishimoto/2/02-087/00001.jpg.

4. Print #196 in Miyata and Takada, *Namazue,* 357. To view this print, see http://gazo.dl.itc.u-tokyo.ac.jp/ishimoto/2/02-039/00001.jpg.

5. In the aftermath of the Great Kantō Earthquake of September 1923, for example, General Ugaki Kazushige characterized the disaster as a "divine punishment against us who had aspired to a culture of materialism and degrading thoughts," and the November 11, 1923, Imperial Rescript Enjoining Sincere and Strenuous Life amplified such sentiments to encourage a greater focus on the public good in citizens' lives. See J. Charles Schencking, "The Great Kantō Earthquake and the Culture of Catastrophe and Reconstruction in 1920s Japan," *Journal of Japanese Studies* 34, no. 2 (summer 2008): 305, 310.

230 NOTES TO PAGES 142-148

6. See, for example, "Tōbu jishin no ki," in *NJS,* vol. 5, supplement 2, part 1, 559; "Ansei daijishin jikkendan," in *NJS,* vol. 5, supplement 2, part 1, 474; and "Yabure mado no ki," in *DNJS,* vol. 1 (*kō*), 560.

7. *FN,* 577. See also *NJS,* vol. 5, supplement 2, part 1, 405 (truncated version), and Wakamizu Suguru, *Edokko kishitsu to namazue* (Kadokawa gakugei shuppan, 2007), 119-125.

8. *FN,* 517.

9. Wakamizu, *Edokko,* 125.

10. Print #14 in Miyata and Takada, *Namazue,* 139, 245-246. See also print #201 (especially the thirteenth verse), 360-361, and Wakamizu, *Edokko,* 112-125. To view print #14, see http://dl.ndl.go.jp/info:ndljp/pid/1301829.

11. Print #43 in Miyata and Takada, *Namazue,* 109, 265-266. See also Abe Yasunari, "Namazue no ue no Amaterasu," *Shisō* 912 (June 2000): 43-44. To view the print, see figure 9 in the present volume.

12. For more details on this matter, see Gregory Smits, "Conduits of Power: What the Origins of Japan's Earthquake Catfish Reveal about Religious Geography," *Japan Review* 24 (2012): 54-65.

13. *DNJS,* vol. 2 (*otsu*), 539.

14. *FN,* 516-517.

15. *FN,* 232-233. See also Noguchi Takehiko, *Ansei Edo jishin: Saigai to seiji kenryōku* (Chikuma shobō, 1997), 229.

16. For a trenchant analysis of this matter and the broader circumstances in East Asia that contributed to it, see Bob Tadashi Wakabayashi, *Anti-Foreignism and Western Learning in Early-Modern Japan: The New Theses of 1825* (Cambridge, MA: Council on East Asian Studies, Harvard University, 1986), 15-16, 61-68.

17. *FN,* 519-520. See also Kitahara Itoko, *Jishin no shakaishi: Ansei daijishin to minshū* (Kōdansha, 2000), 136-137.

18. Constantine Nomikos Vaporis, *Tour of Duty: Samurai, Military Service in Edo, and the Culture of Early Modern Japan* (Honolulu: University of Hawai'i Press, 2008), 2.

19. Nagura Tetsuzō, *Fūshigan ishin henkaku: Minshū wa tennō o dō miteita ka* (Kōsō shobō, 2004), 189-191.

20. Quoted in Nagura, *Fūshigan ishin henkaku,* 193. The verse is loosely based on *Haykunin isshu* 17 (http://etext.lib.virginia.edu/japanese/hyakunin/hyakua.html).

21. Quoted in Nagura, *Fūshigan ishin henkaku,* 180.

22. Ibid., 185.

23. Ibid., 165.

24. Ibid., 186.

25. Nagura, *Fūshigan ishin henkaku,* 161–163.
26. For example, in 1862 broadside prints criticized the emperor for sacrificing his daughter, Princess Kazunomiya, to political expediency. See Nagura Tetsuzō, *Etoki bakumatsu fūshiga to tennō* (Kashiwa shobō, 2007), 14–15.
27. *AKS,* vol. 1, 1. See also Arakawa Hidetoshi, ed., *Jitsuroku, Ō-Edo kaimetsu no hi: Ansei kenmonroku, Ansei kenmonshi, Ansei fūbunshū* (Kyōikusha, 1982), 99.
28. *FN,* 536.
29. Print #142 in Miyata and Takada, *Namazue,* 236, 327–328. To view this image, see http://gazo.dl.itc.u-tokyo.ac.jp/ishimoto/2/02–010/00001.jpg.
30. Translation of the text of this print is found in Peter Duus, *The Japanese Discovery of America: A Brief History with Documents* (New York: Bedford Books, 1997), 110, 112, and in M. William Steele, *Alternative Narratives in Japanese History* (New York: RoutledgeCurzon, 2003), 16–17.
31. Abe, "Amaterasu," 32–35.
32. For example, see #126 and #127 in Miyata and Takada, *Namazue,* 110, 231, 319–320.
33. Print #131 in Miyata and Takada, *Namazue,* 8, 321. To view this print, see Figure 10.
34. Abe, "Amaterasu," 29–32. Further support for the possible reading of "great country" as Japan would be that the text of several other prints with no connection to the black ships or foreigners, begins with "The soil of the great country moves" (*daikoku no tsuchi* . . .), a verbatim match with the song. See, for example, print #191 in Miyata and Takada, *Namazue,* 355.
35. Kitai Toshio, *Shinkokuron no keifu* (Hōzōkan, 2006), 7–17 and Satō Hiroo, *Shinkoku Nihon,* Chikuma shinsho 591 (Chikuma shobō, 2006), 21–41.
36. From *Konjaku monogatarishū,* quoted in Kitai, *Shinkokuron,* 20.
37. The situation in medieval times was more complex and varied than this brief summary can indicate. For the full picture, see Kitani, *Shinkokuron,* 18–78 (esp. the typology of *shinkoku* meanings, pp. 16–25) and Satō, *Shinkoku,* 58–120.
38. Kitani, *Shinkokuron,* 109–117, quoted passage, 112.
39. Kitani, *Shinkokuron,* 117–180.
40. Anna Beerens, *Friends, Acquaintances, Pupils and Patrons, Japanese Intellectual Life in the Late Eighteenth Century: A Prosopographical Approach* (Leiden: Leiden University Press, 2006).
41. Wakabayashi, *Anti-Foreignism,* 39.
42. Wakabayashi, *Anti-Foreignism,* 125.
43. Wakabayashi, *Anti-Foreignism,* 149.
44. Print #68 in Miyata and Takada, *Namazue,* 111, 284–285.

45. AKR, vol. 3, 13–16 (ge no jūsan–ge no jūroku). See also Arakawa *Jitsuroku, Ō-Edo kaimetsu no hi,* 90–94 and Kitahara, *Jishin no shakaishi,* 188. See also Abe, "Amaterasu," 37–38.

46. Print #30 in Miyata and Takada, *Namazue,* 112, 257.

47. Print #31 in Miyata and Takada, *Namazue,* 112, 258.

48. Kitai, *Shinkokuron,* 118–120. Izumo, Nara and Yamato provinces were also sometimes referred to as *shinkoku* in the context of Warring States daimyō encroaching on these territories, which had been under control of shrines or temples. Once again, the term *shinkoku* indicated a crisis that threatened the social structure, albeit at the local level in these cases.

49. Print #44 in Miyata and Takada, *Namazue,* 106, 266. In this capacity, Ebisu was acting as a *rusu(i)gami,* that is, a caretaker deity who stands in for the main *kami* during the tenth month. Although often depicted as good-natured, Ebisu's origins are complex and contain a dark side that can manifest itself in certain circumstances. For a detailed explanation of these matters, see Cornellis Ouwehand, *Namazu-e and their Themes: An Interpretative Approach to Some Aspects of Japanese Folk Religion* (Leiden: E. J. Brill, 1964), 16, 82–85. To view print #44, see http://metro2.tokyo.opac.jp/tml/tpic/imagedata/ toritsu/ukiyoe/0C/0277-C040.jpg.

50. Wakamizu Suguru, *Namazu wa odoru: Edo no namazue omoshiro bunseki* (Bungeisha, 2003), 70.

51. Noguchi, *Ansei Edo jishin,* 194–195.

52. Image #29 in Miyata and Takada, *Namazue,* 257. For analysis of this image, see Abe, "Amaterasu," 40.

53. Image #32 in Miyata and Takada, *Namazue,* 258–60.

54. Abe, "Amaterasu," 41–42.

55. For example, see prints #66 and #68 in Miyata and Takada, *Namazue,* 283, 285.

56. T. Fujitani, *Splendid Monarchy: Power and Pageantry in Modern Japan* (Berkeley: University of California Press, 1996), 1–9.

57. Gregory Clancey, *Earthquake Nation: The Cultural Politics of Japanese Seismicity, 1868–1930* (Berkeley: University of California Press, 2006), 131.

58. Print #191 in Miyata and Takada, *Namazue,* 17, 355. To view this print, see http://metro2.tokyo.opac.jp/tml/tpic/imagedata/toritsu/ ukiyoe/0C/0277-C031.jpg.

59. Noguchi, *Ansei Edo jishin,* 212–216. Noguchi points out that this disk represents the sun, not the moon as some have speculated. A rabbit in a disk indeed typically signifies the moon, but the color of the disk in this print is red, although washed out in many copies.

60. Miyazaki Narumi points out that talk of Enma no ko (child of hell-king Yama) appeared the previous year in Edo. The child was a miniature version of his

father and very strong. According to hearsay, people trapped under beams would sometimes call out for Enma no ko to come and help them. "Ansei itsubō jishin kibun," in *NJS*, vol. 5, supplement 2, part 1, 450. In a catfish print, *Enma no ko,* earthquake victims in Yama's court beg to be treated as his child. Print #16 in Miyata and Takada, *Namazue*, 246–247. To view this print, see http://gazo.dl.itc.u-tokyo.ac.jp/ishimoto/2/02–083/00001.jpg.

61. Anne Walthall, "Edo Riots," in James L. McClain, John W. Merriman, and Ugawa Kaoru, eds., *Edo and Paris: Urban Life and the State in the Early Modern Era* (Ithaca, NY: Cornell University Press, 1994), 415–416.

62. Takagi Shunsuke, *Ee ja nai ka* (Kyōikusha, 1979), 208–231.

63. Ibid., 222–224.

64. Ibid., 225–227. The name comes from the apocryphal tales featuring Chōshū loyalists shouting "Zannen!" (Too bad!) just before they died.

65. *FN*, 554. See also Noguchi, *Ansei Edo jishin,* 31–32. The name Sōgorō is surely significant, owing to the fame of Sakura Sōgorō, a seventeenth-century martyr who by the nineteenth century "had become the patron saint of protest." Moreover, Sōgorō had become especially popular in Edo after his story appeared on the Kabuki stage in 1851. See Anne Walthall, ed., trans., *Peasant Uprisings in Japan* (Chicago: University of Chicago Press, 1991), 35–75, esp. 37.

66. Stephen Vlastos, *Peasant Protests and Uprisings in Tokugawa Japan* (Berkeley: University of California Press, 1986), 153.

67. When the Tokugawa period ended, so did these traditional restraints on violence. Early Meiji period social protests often turned deadly. See, for example, David L. Howell, *Geographies of Identity in Nineteenth-Century Japan* (Berkeley: University of California Press, 2005), esp. 89–109.

68. Print #19 in Miyata and Takada, *Namazue*, 248–249.

69. Howell, *Geographies of Identity,* 100–101.

70. Noguchi, *Ansei Edo jishin,* 18.

71. Print #45 in Miyata and Takada, *Namazue*, 6–7, 266–267. This print is also known as *Mizugami no tsuge.* To view the relevant half of it, see http://metro2 .tokyo.opac.jp/tml/tpic/imagedata/toritsu/ukiyoe/0C/0277-C004(02).jpg.

72. Several catfish prints feature groups of small catfish representing recent past earthquakes. Some go back only to Zenkōji, and others include Zenkōji but go back as far as Kyoto. See, for example, prints #37 and #76 in Miyata and Takada, *Namazue*, 110, 262, 292.

73. *AKS*, vol. 3, 15. See also Arakawa *Jitsuroku, Ō-Edo kaimetsu no hi,* 183, 186–187, and Kitahara, *Jishin no shakaishi,* 173–176.

74. "Gojōsho," in *DNJS*, vol. 2 (*otsu*), 549.

75. Wakamizu, *Edokko kishitsu,* 6.

76. Noguchi, *Ansei Edo jishin,* 216–222.

234 NOTES TO PAGES 169–176

77. *FN,* 518.
78. Julia Adeney Thomas, *Reconfiguring Modernity: Concepts of Nature in Japanese Political Ideology* (Berkeley: University of California Press, 2001), 38.
79. Thomas, *Reconfiguring Modernity,* 43.
80. George M. Wilson, *Patriots and Redeemers in Japan: Motives in the Meiji Restoration* (Chicago: University of Chicago Press, 1982), 78.

Chapter 6   Into the Twenty-First Century

1. Jelle Zeilinga de Boer and Donald Theodore Sanders, *Earthquakes in Human History: The Far-Reaching Effects of Seismic Disruptions* (Princeton, NJ: Princeton University Press, 2005), x–xi.
2. Rachel Laudan, *From Mineralogy to Geology: The Foundations of a Science, 1650–1830* (Chicago: University of Chicago Press, 1987), 43.
3. Laudan, *Mineralogy to Geology,* 218.
4. Charles Lyell, *Principles of Geology: Being an Attempt to Explain the Former Changes of the Earth's Surface, by Reference to Causes Now in Operation,* 2nd ed., vol. 1 (London: John Murray, 1832), 539–540. For the complete discussion of earthquakes, volcanoes, and related phenomena, see 457–553. For theories of earthquakes, see 533–545.
5. Ibid., 543.
6. Ibid., 505.
7. Susan Elizabeth Hough, *Earthshaking Science: What We Know (and Don't Know) about Earthquakes* (Princeton, NJ: Princeton University Press, 2002), 205.
8. Kotô, Bundjiro [Kotō Bunjirō], "On the Cause of the Great Earthquake in Central Japan, 1891," *Tōkyō teikoku daigaku kiyō, rika* 5, no. 10 (1893): 295–353.
9. Tsuji Yoshinobu, *Sennen shinsai: Kurikaesu jishin to tsunami no rekishi ni manabu* (Daiyamondo sha, 2011), 252–253.
10. Peter M. Shearer, *Introduction to Seismology,* 2nd ed. (New York: Cambridge University Press, 2009), 2.
11. Gregory Clancey, *Earthquake Nation: The Cultural Politics of Japanese Seismicity, 1868–1930 (*Berkeley: University of California Press, 2006), esp. 63–66. See also Boumsoung Kim, *Meiji, Taishō no Nihon no jishingaku: Rokaru saiensu o koete* (Tōkyō daigaku shuppankai, 2007), 19–48.
12. For more details, see Gregory Smits, "Danger in the Lowground: Historical Context for the March 11, 2011 Tōhoku Earthquake and Tsunami," *Asia-Pacific Journal* 9, issue 20, no. 4 (May 16, 2011), at http://www.japanfocus .org/-Gregory-Smits/3531.
13. Iki Tsunenaka, "Sanriku chihō tsunami jistujō torishirabe hōkoku," in *Shinsai yobō chōsakai hōkoku, dai 7 gō* (Shinsai yobō chōsakai, 1896), 30–33.
14. Susanna M. Hoffman, "The Monster and the Mother: The Symbolism of Disaster," in Susanna M. Hoffman and Anthony Oliver-Smith, eds.,

Catastrophe and Culture: The Anthropology of Disaster (Santa Fe, NM: School of American Research Press, 2002), 137–139.

15. Ibid., 114. Hoffman points out that one message tied around a tree in the wake of the Oakland firestorm of 1991 read, "May We All Be Restored and Renewed" (116).

16. *Yomiuri shinbun,* September 6, 1877, morning edition, 4.

17. "Ōsegaki," *Yomiuri shinbun,* November 26, 1887, morning edition, 2.

18. "Daishinsai no yonjūichi shūki," *Yomiuri shinbun,* October 31, 1894, morning edition, 5.

19. "Ansei daijishin no daikuyō," *Yomiuri shinbun,* March 27, 1903, morning edition, 4.

20. "Jishin gakkai," *Yomiuri shinbun,* May 31, 1888, morning edition, 2.

21. "Jishin to arashi," *Yomiuri shinbun,* January 4, 1885, special edition, 3. Lyell also commented on earthquakes and weather: "That there is an intimate connexion between subterranean convulsions and particular states of the weather is unquestionable; but, as Michell truly remarked, 'it is more probably that the air should be affected by the causes of earthquakes, than that the earth should be affected in so extraordinary a manner, and to so great a depth, by a cause residing in the air.'" Lyell, *Principles of Geology,* vol. 1, 534. The focus of Japanese interest in atmospheric phenomena tended not to be concerned with the possibility that certain types of weather might cause earthquakes or vice versa but with the possibility that certain types of weather might indicate an earthquake will soon strike.

22. Tomiharu Isao, "Sekiyū to jishin to no kankei," *Yomiuri shinbun,* February 24, 1885, morning edition, 1.

23. "Jishin o kizukau mono ari," *Yomiuru shinbun,* November 4, 1891, special edition, 2.

24. Although Musha Kinkichi disagreed with the notion that unseasonably warm weather was a sign of an impending earthquake, he related a tale from the Ansei Edo earthquake of a bannerman on guard duty who announced that an earthquake would soon occur, and emergency preparations were made in time. When later asked the reason for his insight, he said that survivors of the Sanjō earthquake reported that the sky appeared closer, the stars shone much more brightly than usual, and it was warm in the winter. Moreover, in connection with the Genroku earthquake, an old man named Amano Yagozaemon said that when the stars appear low in the sky and the weather is warm in the winter, an earthquake is on the way. He reinforced his house before the shaking started. Musha Kinkichi, *Jishin namazu* (Meiseki shoten, 1995, originally 1957), 133.

25. To cite but a few of the almost endless examples, Ikeya Motoji frequently quotes "old sayings" or "proverbs" such as "a yellow morning sun, a red moon, and twinkling stars are precursors" or "when wind blows up from the ground

there will be an earthquake." Significantly, Motoji never mentions the historical context of these sayings, namely the Tokugawa-era notions about trapped yang energy within the earth. See *Earthquakes and Animals: From Folk Legends to Science* (River Edge, NJ: World Scientific, 2004), 7, 8. Benjamin Reilly revives the theory of meteorologist C. F. Brooks that "cyclonic weather" was a significant causal factor in the 1923 Great Kantō Earthquake. See *Disaster and Human History: Case Studies in Nature, Society, and Catastrophe* (Jefferson, NC: McFarland & Company, 2009), 82. As discussed in the introduction, many of the articles in Saumitra Mukherjee, ed., *Earthquake Prediction* (Leiden: Brill, 2006) posit atmospheric earthquake causes or precursors.

26. For a discussion of this post-Nōbi turn toward science, see Clancey, *Earthquake Nation,* 151–179.

27. "Kinō no daijishin," *Yomiuri shinbun,* June 21, 1894, morning edition, 2.

28. "Jishin no chūi," *Yomiuri shinbun,* January 16, 1915, morning edition, 5.

29. "Tsūzoku jishin monogatari (*jō*)," *Yomiuri shinbun,* November 19, 1915, morning edition, 5.

30. "Tōkyō to jishin: shinpai no oyobazu," *Yomiuri shinbun,* October 18, 1917, morning edition, 5.

31. Clancey, *Earthquake Nation,* 220–226.

32. "Jishin kasai no tenrankai," *Yomiuri shinbun,* November 18, 1923, morning edition, 5.

33. In *Namazu Taiheiki konzatsubanashi,* for example, Kashima's victorious forces sell the defeated catfish rebels to local restaurants. See *FN,* 519. The theme also occurs frequently in catfish prints: for example, prints #57, #58, #59, and #60, Miyata and Takada, *Namazue,* 228–229, 274–278.

34. "Ansei daijishin no komonjo hakken," *Yomiuri shinbun,* September 20, 1953, morning edition, 6.

35. See Clancey, *Earthquake Nation,* 221.

36. For a list of precursors (*zenchō*) in one account, see "Ōjishin ōkaze kenmonki," in *NJS,* vol. 5, supplement 2, part 1, 537–538.

37. Clancey, *Earthquake Nation,* 152–153.

38. "Tsūzoku jishin monogatari (*jō*)," *Yomiuri shinbun,* November 19, 1915, morning edition, 5.

39. According to seismologist Shimamura Hideki, contemporary science has been unable to shed any light on a possible relationship between volcanoes and earthquakes. See *Nihonjin ga shiritai jishin no gimon 66: Jishin ga ōi Nihon dakarakoso chishiki no sonae mo wasurezuni!* (Sofutobanku kurieitibu, 2008), 90–92.

40. Musha, *Jishin namazu,* 150.

41. Thomas Gilovich, *How We Know What Isn't So: The Fallibility of Human Reason in Everyday Life* (New York: Free Press, 1991, 1993), 50.

42. Gilovich, *How We Know What Isn't So*, 72, and Shimamura, *Jishin no gimon*, 102–103. For a detailed analysis of this and related psychological phenomena, in addition to Gilovich see Kikuchi Satoru, *Chōetsu genshō o naze shinjiru no ka: omoikomi o umu "taiken" no ayausa* (Kōdansha, 1998).

43. Regarding sensitivity to electricity, Shimamura points out that even if catfish can sense some kind of small change that might occur prior to an earthquake, "civilization" in the form of electricity use has greatly increased background electrical noise. Geophysical sensors located even several kilometers from railway lines, for example, pick up considerable electrical noise. It is doubtful that catfish would be able to detect subtle electrical fluctuations in such a "noisy" environment. Shimamura, *Jishin no gimon*, 105–106.

44. "Namazu ga ugoku to kanarazu jishin ga okoru," in *Yomiuri shinbun*, April 1, 1932, morning edition, 7. See also excerpts of a presentation Hatai gave on his research, "Namazu no ugoki de jishin o yochi," in *Yomiuri shinbun*, October 14, 1932, morning edition, 4.

45. Robert Geller (Robaato Geraa), *Nihonjin wa shiranai "Jishin yochi" no shōtai* (Futabasha, 2011), 133–134.

46. For full details on this research, see Rikitake Tsuneji, "Deeta ni miru namazu to jishin," in Miyata and Takada, *Namazue*, 148–156. For details on research in the 1920s, see Musha, *Jishin namazu*, 16–22. Regarding animals other than fish, see 23–51.

47. Quoted in Geller, *Jishin yochi*, 134–135. Geller points out critically that if one in a hundred is a suitable yardstick, then every superstition or fantasy ever connected with earthquakes deserves public funding.

48. Musha, *Jishin namazu*, 19–21. More recently, Motoji and many others have made the same claim. See *Earthquakes and Animals* for abundant examples.

49. Musha, *Jishin namazu*, 23–24.

50. Ibid., 24–29, and Yasuo Suyehiro [Suehiro], "Some Observations on the Unusual Behavior of Fishes Prior to an Earthquake," Earthquake Research Institute, Tokyo Imperial University (March 1934), 228–231. According to Suehiro, recent studies "equally prove the existence of a mysterious and instinctive reaction of the fish to earthquakes" (228). For photos of sardines and the threadfish, see 230.

51. Quoted in Geller, *Jishin yochi*, 133.

52. Musha, *Jishin namazu*, 31–35. See also Yoshimura Akira, *Sanriku kaigan ōtsunami* (Bungei shunjū, 2004), 16–20, 81–82.

53. Musha, *Jishin namazu*, 36–40.

54. Ibid., 40. It remains a common rhetorical technique among advocates of earthquake precursors and prediction to claim that they are offering possibilities, not definite proof or precise tools. Motoji, for example, says, "I am not arguing for earthquake prediction using the animal precursors . . .

if, by prediction, we are meaning an exact forecast of time, epicenter, and magnitude of an earthquake. In that case no prediction is possible for any fracture phenomena." See *Earthquakes and Animals,* ix.

55. Gilovich, *How We Know What Isn't So,* 58.

56. Musha, *Jishin namazu,* 40–50.

57. Ibid., 51.

58. Gilovich, *How We Know What Isn't So,* 158.

59. This topic is beyond the scope of the present study. For a thorough analysis, see Geller, *Jishin yochi,* esp. 76–159, and Shimamura Hideki, *"Jishin yochi" wa uso darake* (Kōdansha, 2008).

60. Clancey, *Earthquake Nation,* quoted passages, 85.

61. Quoted in Julia Adeney Thomas, *Reconfiguring Modernity: Concepts of Nature in Japanese Political Ideology* (Berkeley: University of California Press, 2001), 172–173.

62. Clancey, *Earthquake Nation,* 106–107.

63. Clancey discusses the role of Italy at length. See especially *Earthquake Nation,* 106–112.

64. Borland, "Capitalising on Catastrophe," 892. For more details on the production of educational materials from the earthquake, see 893–905, and Janet Borland, "Stories of Ideal Japanese Subjects from the Great Kanto Earthquake of 1923," *Japanese Studies* 25, no. 1 (May 2005): 21–34.

65. Borland, "Capitalising on Catastrophe," 893–905.

66. Timothy Yun Hui Tsu, "Making Virtues of Disaster: 'Beautiful Tales' from the Kobe Flood of 1938," *Asian Studies Review* 32, no. 3 (June 2008): 197.

67. Tsu, "Making Virtues of Disaster," 198.

68. Yamashita Funio, *Tsunami no kyōfu: Sanriku tsunami denshōroku* (Sendai, Japan: Tōhoku daigaku shuppankai, 2005), 99.

69. Geller, *Jishin yochi,* 24–66.

70. The coverage of this matter in the newspapers has been extensive. For example, see "Higashi Nihon dai-shinsai: Kyodai tsunami rokusen-nen de rokkai, chisō ni konseki," in *Mainichi shinbun,* August 21, 2011, and "Q: Sen-nen ni ichido no kyodaijishin to iwareru Higashi Nihon daishinsai, naze sen-nen ni ichihido na no? A: Jōgan jishin no sairai toiu mikata," in *Yomiuri Online,* October 20, 2011. See also Dengler and Smits (introduction), "The Past Matters," for some discussion of paleoseismology.

71. Tsuji, *Sennen shinsai.*

BIBLIOGRAPHY OF WORKS CITED

Abbreviations of Frequently Cited Primary Sources or Collections

AKR. Ansei kenmonroku. Hattori Yasunari (text) and Utagawa Yoshiharu (illustrations).

AKS. Ansei kenmonshi. Illustrations by Utagawa Kuniyoshi et al.

DNJS. Dai-Nihon jishin shiryō. Shinsai yobō chōsakai, eds.

FN. Fujiokaya nikki. Suzuki Tōzō and Koike Shōtarō, eds.

NJS. Shinshū nihon jishin shiryō. Tōkyō daigaku jishin kenkyūjo, ed.

NRJSS. "Nihon no rekishi jishin shiryō" shūi. Usami Tatsuo, ed.

SGS. Ansei Edo jishin saigaishi. Sayama Mamoru.

Note: The place of publication for Japanese books is Tokyo unless otherwise indicated.

Abe Yasunari. "Jishin to hitobito no sōzōryoku." In Chūō bōsai kaigi, eds., *1855 Ansei Edo jishin hōkokusho.* Fuji sōgō kenkyūsho, 2004, 129–200.

———. "Namazue no ue no Amaterasu." In *Shisō* 912 (June 2000), 25–52.

Akahane Sadayuki. "Zenkōji jishin to saigai no zenbō." In Akahane Sadayuki and Kitahara Itoko, eds., *Zenkōji jishin ni manabu.* Nagano-shi: Shinano mainichi shinbunsha, 2003, 9–34.

Akahane Sadayuki and Kitahara Itoko, eds. *Zenkōji jishin ni manabu.* Nagano-shi: Shinano mainichi shinbunsha, 2003.

Akasoy, Anna A. "Interpreting Earthquakes in Medieval Islamic Texts." In Christof Mauch and Christian Pfister, eds., *Natural Disasters, Cultural Responses: Case Studies toward a Global Environmental History.* New York: Lexington Books, 2009, 183–196.

Akira Naito. *Edo, the City That Became Tokyo: An Illustrated History.* Illustrations by Kazuo Hozumi. Translated, adapted, and introduced by H. Mack Horton. Kodansha, 2003.

Alabaster, Jay. "Tsunami-Hit Towns Forgot Warnings from Ancestors." Associated Press, April 6, 2011.

Ansei kenmonshi, 3 vols. Illustrations by Utagawa Kuniyoshi et al. Author(s) and publisher unknown, 1856.

Aoki Kunio et al., eds. *Taikyoku jishinki, Ansei kenmonroku, Jishin yobōsetsu, Bōkasaku zusetsu.* Edo kagaku koten sōsho 19. Kōwa shuppan, 1979.

Arakawa Hidetoshi, ed. *Jitsuroku, Ō-Edo kaimetsu no hi: Ansei kenmonroku, Ansei kenmonshi, Ansei fūbunshū.* Kyōikusha, 1982.

Asai Ryōi. *Kaname'ishi* (1662). In Taniwaki Masachika, Oka Masahiko, and Inoue Kazuhito, eds., trans., *Kanazōshishū*. Shōgakkan, 1999, 13–83.

Aston, W. G., trans. *Nihongi: Chronicles of Japan from the Earliest Times to A.D. 697*. 2 vols. Rutland, VT: Charles E. Tuttle Co., 1972.

Azuma Kagami (ca. 1266). http://www5a.biglobe.ne.jp/~micro-8/toshio/azuma/123105.html. Accessed February 14, 2011.

Bakun, William H. "Magnitude and Location of Historical Earthquakes in Japan and Implications for the 1855 Ansei Edo Earthquake." *Journal of Geophysical Research* 110, B02304 (2005), 1–22.

Beerens, Anna. *Friends, Acquaintances, Pupils and Patrons, Japanese Intellectual Life in the Late Eighteenth Century: A Prosopographical Approach*. Leiden: Leiden University Press, 2006.

Berry, Mary Elizabeth. *Japan in Print: Information and Nation in the Early Modern Period*. Berkeley: University of California Press, 2006.

Bix, Herbert P. *Peasant Protest in Japan, 1590–1884*. New Haven, CT: Yale University Press, 1986.

Bolitho, Harold. "The Tempō Crisis." In Marius B. Jansen, ed., *The Cambridge History of Japan*, vol. 5, *The Nineteenth Century*. New York: Cambridge University Press, 1989, 116–167.

Bolt, Bruce A. *Earthquakes*. Newly Revised and Expanded. New York: W. H. Freeman and Company, 1993 (fifth printing, 1997).

Borland, Janet. "Capitalising on Catastrophe: Reinvigorating the Japanese State with Moral Values through Education following the 1923 Great Kantō Earthquake." *Modern Asian Studies* 40, no. 4 (October 2006): 875–907.

———. "Stories of Ideal Japanese Subjects from the Great Kanto Earthquake of 1923." *Japanese Studies* 25, no. 1 (May 2005): 21–34.

Buyō Inshi. *Seji kenmonroku*. Honjō Eijirō, ed. Seiabō, 2001 (originally published 1816).

Chen Fu. *Nongshu*. Taibei: Taiwan shangwu yinshuguan, 1956, 2 vols. (originally published 1149).

Chūō bōsai kaigi. *1662 Kanbun Ōmi-Wakasa jishin hōkokusho*. Mizuho jōhō sōken kabushikigaisha, 2005.

———. *1847 Zenkōji jishin hōkokusho*. Nihon shisutemu kaihatsu kenkyūsho, 2007.

———. *1854 Ansei Tōkai jishin, Ansei Nankai jishin hōkokusho*. 2005.

———. *1855 Ansei Edo jishin hōkokusho*. Fuji sōgō kenkyūsho, 2004.

———. *1891 Nōbi jishin hōkokusho*. Nihon shisutemu kaihatsu kenkyūsho, 2008.

———. *1896 Meiji Sanriku jishin tsunami hōkokusho*, 2005.

———. *1923 Kantō daishinsai hōkokusho, dai-ippen*. Nihon shisutemu kaihatsu kenkyūsho, 2006.

"City Looks to Base Tsunami Warnings on Animal Behavior." *Japan Times,*
June 3, 2012.

Clancey, Gregory. *Earthquake Nation: The Cultural Politics of Japanese Seismicity,*
1868–1930. Berkeley: University of California Press, 2006.

Duus, Peter. *The Japanese Discovery of America: A Brief History with Documents.*
New York: Bedford Books, 1997.

Edo ōjishin kiji. 1855. (Scroll consisting of earthquake materials).

Ekken-kai, eds. *Ekken zenshū,* vol. 3. Ekken zenshū kankōbu, 1911.

Elisonas, Jurgis. "Notorious Places: A Brief Excursion into the Narrative
Topography of Early Edo." In James L. McClain, John W. Merriman, and
Ugawa Kaoru, eds., *Edo and Paris: Urban Life and the State in the Early*
Modern Era. Ithaca, NY: Cornell University Press, 1994, 253–291.

Formanek, Susanne, and Sepp Linhart. *Written Texts—Visual Texts: Woodblock-*
Printed Media in Early Modern Japan. Amsterdam: Hotei Publishing, 2005.

Fujita Rihei. *Edo kanoko.* Asakura Haruhiko, ed. Sumiya shobō, 1970
(originally 1687).

Fujitani, Takashi. *Splendid Monarchy: Power and Pageantry in Modern Japan.*
Berkeley: University of California Press, 1996.

Fulton, Elaine. "Acts of God: The Confessionalization of Disaster in Reformation
Europe." In Andrea Janku, Gerrit Schenk, and Franz Mauelshagen, eds.,
Historical Disasters in Context: Science, Religion, and Politics. New York:
Routledge, 2012, 54–74.

Furihata Hiroki. "Zenkōji jishin to saigai jōhō." In Akahane Sadayuki and
Kitahara Itoko, eds., *Zenkōji jishin ni manabu.* Nagano-shi: Shinano mainichi
shinbunsha, 2003, 139–165.

Geller, Robert (Robaato Geraa). *Nihonjin wa shiranai "jishin yochi" no shōtai.*
Futabasha, 2012.

Gilovich, Thomas. *How We Know What Isn't So: The Fallibility of Human Reason in*
Everyday Life. New York: Free Press, 1991, 1993.

Gotō Yōichi and Tomoeda Ryūtarō, eds. *Kumazawa Banzan,* Nihon shisō taikei
30. Iwanami shoten, 1971.

Gramlich-Oka, Bettina, and Gregory Smits, eds. *Economic Thought in Early*
Modern Japan. Leiden: Brill, 2010.

Groemer, Gerald. *Bakumatsu no hayari uta: Kudokibushi to bushi no shin kenkyū.*
Meicho shuppan, 1995.

———. "Singing the News: Yomiuri in Japan during the Edo and Meiji Periods."
Harvard Journal of Asiatic Studies 54, no. 1 (June 1994): 233–261.

Guidoboni, Emanuela, and John E. Ebel. *Earthquakes and Tsunamis in the Past:*
A Guide to Techniques in Historical Seismology. New York: Cambridge
University Press, 2009.

Hagiwara Takahiro et al. *Kojishin: Rekishi shiryō to katsudansō kara saguru.* Tōkyō daigaku shuppankai, 1982.

Harada Kazuhiko, "Matsushiro-han de sakusei sareta jishin zuerui ni tsuite." In Chūō bōsai kaigi, eds., *1847 Zenkōji jishin hōkokusho.* Nihon shisutemu kaihatsu kenkyūsho, 2007: [156].

Harley, J. B., and David Woodward, eds. *The History of Cartography,* vol. 2, book 2: *Cartography in the Traditional East and Southeast Asian Societies.* Chicago: University of Chicago Press, 1994.

Hashimoto Manpei. *Jishingaku kotohajime: Kaitakusha Sekiya Seikei no shōgai.* Asahi shinbunsha, 1983.

Hata Ginkei. *Shigure no sode.* In Edo sōsho kankōkai, ed., *Edo sōsho,* vol. 10. Meicho kankōkai, 1961, 1–306.

Hatano, Jun. "Edo's Water Supply." In James L. McClain, John W. Merriman, and Ugawa Kaoru, eds., *Edo and Paris: Urban Life and the State in the Early Modern Era.* Ithaca, NY: Cornell University Press, 1994, 234–250.

Hattori Yasunari (text), and Utagawa Yoshiharu (illustrations). *Ansei kenmonroku,* 3 vols. [publisher unknown], 1856. Pagination visible, with each full sheet, folded into two page faces.

Hayashi, Reiko. "Provisioning Edo in the Early Eighteenth Century: The Pricing Policies of the Shogunate and the Crisis of 1733." In James L. McClain, John W. Merriman, and Ugawa Kaoru, eds., *Edo and Paris: Urban Life and the State in the Early Modern Era.* Ithaca, NY: Cornell University Press, 1994, 211–233.

"Higashi Nihon dai-shinsai: Kyodai tsunami rokusen-nen de rokkai, chisō ni konseki." In *Mainichi shinbun,* August 21, 2011.

Higashi Sachiyo. "Kanbun jishin ga motarashita mono." In Chūō bōsai kaigi, *1662 Kanbun Ōmi-Wakasa jishin hōkokusho.* Mizuho jōhō sōken kabushikigaisha, 2005, 119–121.

———. "Uramikawa kussaku jigyō to shinden kaihatsu." In Chūō bōsai kaigi, *1662 Kanbun Ōmi-Wakasa jishin hōkokusho.* Mizuho jōhōsōken kabushikigaisha, 2005, 115–119.

Hoffman, Susanna M. "The Monster and the Mother: The Symbolism of Disaster." In Susanna M. Hoffman and Anthony Oliver-Smith, eds., *Catastrophe and Culture: The Anthropology of Disaster.* Santa Fe, NM: School of American Research Press, 2002, 113–141.

Hoffman, Susanna M., and Anthony Oliver-Smith, eds. *Catastrophe and Culture: The Anthropology of Disaster.* Santa Fe, NM: School of American Research Press, 2002.

Hough, Susan Elizabeth. *Earthshaking Science: What We Know (and Don't Know) about Earthquakes.* Princeton, NJ: Princeton University Press, 2002.

———. *Predicting the Unpredictable: The Tumultuous Science of Earthquake Prediction.* Princeton, NJ: Princeton University Press, 2010.

Howell, David L. *Geographies of Identity in Nineteenth-Century Japan.* Berkeley: University of California Press, 2005.

Ikeya, Motoji. *Earthquakes and Animals: From Folk Legends to Science.* River Edge, NJ: World Scientific, 2004.

Iki Tsunenaka. "Sanriku chihō tsunami jistujō torishirabe hōkoku." In *Shinsai yobō chōsakai hōkoku, dai 7 gō* (Shinsai yobō chōsakai, 1896), 4–34. Available at http://hdl.handle.net/2261/16743.

Imahori Ta'itsu. "Kokudo no saigai to akukishin: Saigai to zokushin." In Inseiki bunka kenkyūkai, eds., *Seikatsushi* (Inseiki bunka ronshū, vol. 5), 198–232.

Inagaki Funio, ed. *Edo no taihen: Jishin, kaminari, kaji, kaibutsu.* Heibonsha, 1995.

Inoue Uin. "Shōwa hachinen sangatsu mikka no jishin ni tomonatta onkyō ni tsuite." Tōkyō teikoku daigaku jishin kenkyūsho, March 1934, 77–86. http://hdl.handle.net/2261/13774 (accessed January 2, 2011).

Ishibashi Katsuhiko. *Daijishinran no jidai.* Iwanami shoten, 1994.

Ishii Ryōsuke. *Edojidai manpitsu,* revised ed., vol. 2. Asahi shinbunsha, 1991.

Itō Kazuaki. *Jishin to funka no Nihonshi.* Iwanami shoten, 2002.

Janku, Andrea, Gerrit Schenk, and Franz Mauelshagen, eds. *Historical Disasters in Context: Science, Religion, and Politics.* New York: Routledge, 2012.

Jansen, Marius B., and Gilbert Rozman, eds. *Japan in Transition: From Tokugawa to Meiji.* Princeton, NJ: Princeton University Press, 1986.

"Japanese City to Watch Animal Behaviour for Disaster Signs." *Tokyo Times,* 2012.

Jishin narabi ni shukka saikenki. Author(s) and publisher unknown, 1855.

Kagan, Yan Y., David G. Jackson, and Robert J. Geller. "Characteristic Earthquake Model, 1884–2011, R.I.P." *Seismological Research Letters* 83 (November/December 2012): 951–953.

Kaibara Ekken. *Kadōkun.* Ekken-kai, eds., *Ekken zenshū,* vol. 3. Ekken zenshū kankōbu, 1911.

Kanagaki Robun. *Ansei fubūnshū,* 3 vols. Publisher unknown, 1856.

Karatani Tomoka. "Kii-koku Hiromura ni tsuite." In Chūō bōsai kaigi, *1854 Ansei Tōkai jishin, Ansei Nankai jishin hōkokusho,* 2005, 67–76.

Kashima Tarō [pseudonym]. *Kashima no kami to Suwa no kami: Higashi Nihon dai shinsai no fukkō ni mukete.* Kashima-shi: Kashima jingū shamusho, 2011.

Katō, Takashi. "Governing Edo." In James L. McClain, John W. Merriman, and Ugawa Kaoru, eds., *Edo and Paris: Urban Life and the State in the Early Modern Era.* Ithaca, NY: Cornell University Press, 1994, 41–67.

Keller, Edward A., and Nicholas Pinter. *Active Tectonics: Earthquakes, Uplift, and Landscape,* 2nd ed. Upper Saddle River, NJ: Prentice Hall, 2002.

Kelly, William W. "Incendiary Actions: Fires and Firefighting in the Shogun's Capital and the People's City." In James L. McClain, John W. Merriman, and Ugawa Kaoru, eds., *Edo and Paris: Urban Life and the State in the Early Modern Era.* Ithaca, NY: Cornell University Press, 1994, 310–331.

Kikuchi Satoru. *Chōetsu genshō o naze shinjiru no ka: Omoikomi o umu "taiken" no ayausa.* Kōdansha, 1998.

Kim, Boumsoung. *Meiji, Taishō no Nihon no jishingaku: Rokaru saiensu o koete.* Tōkyō daigaku shuppankai, 2007.

Kinoshita Naoyuki and Yoshimi Shunya, eds. *Nyūsu no tanjō: Kawaraban to shinbun nishikie no jōhō sekai.* Tōkyō daigaku sōgō kenkyū hakubutsukan, 1999.

Kitahara Itoko. "Ansei daijishin." In Kinoshita Naoyuki and Yoshimi Shunya, eds., *Nyūsu no tanjō: Kawaraban to shinbun nishiki-e no jōhōsekai.* Tōkyō daigaku sōgō kenkyū hakubutsukan, 1999, 154–167.

———. "Ansei tōkai, nankai jishin no higai jōhō ni tsuite: Kawaraban o chūshin ni." In Chūō bōsai kaigi, eds., *1854 Ansei Tōkai jishin, Ansei Nankai jishin hōkokusho,* 2005, 77–89.

———. *Jishin no shakaishi: Ansei daijishin to minshū.* Kōdansha, 2000.

———. "Kinsei ni okeru saigai kyūsai to fukkō." In Kitahara Itoko, ed., *Nihon saigaishi.* Yoshikawa kōbunkan, 2006, 196–229.

———. *Kinsei saigai jōhōron.* Hanawa shobō, 2003.

———, ed., *Nihon saigaishi.* Yoshikawa kōbunkan, 2006.

———. "Saigai to jōhō." In Kitahara Itoko, ed., *Nihon saigaishi.* Yoshikawa kōbunkan, 2006, 230–260.

———. "Saigai to kawaraban: Sono rekishiteki tenkai." In Kinoshita Naoyuki and Yoshimi Shunya, eds., *Nyūsu no tanjō: Kawaraban to shinbun nishiki-e no jōhōsekai.* Tōkyō daigaku sōgō kenkyū hakubutsukan, 1999, 25–43.

———. "Saigai to Shakaizō." In Chūō bōsao kaigi, eds., *1855 Ansei Edo jishin hōkokusho.* Fuji sōgō kenkyūsho, 2004, 43–125.

———. "Shimoda-kō no higai to fukkō." In Chūō bōsai kaigi, eds., *1854 Ansei Tōkai jishin, Ansei Nankai jishin hōkokusho,* 2005, 20–42.

Kitai Toshio. *Shinkokuron no keifu.* Hōzōkan, 2006.

Kitani Makoto. "Kurobune to jishin namazu: Namazue no fūdo to jidai." In Miyata Noboru and Takada Mamoru, eds., *Namazue: Shinsai to Nihon bunka.* Ribun shuppan, 1995, 52–62.

———. "Namazue to yakuharai." Tsuchiura shiritsu hakubutsukan, ed., *Namazue kenbunroku: Ō-Edo bakumatsu namazue jijō.* Tsuchiura-shi: Tsuchiura shiritsu hakubutsukan, 1996, 39–50.

Kitō Yasuyuki and Nagase Satoshi. "Saigai to kyūsai: Machi to mura." In Akahane Sadayuki and Kitahara Itoko, eds., *Zenkōji jishin ni manabu.* Nagano-shi, Japan: Shinano mainichi shinbunsha, 2003, 75–94.

Koga rekishi hakubutsukan, eds. *Tenpenchii to seikimatsu: Nihonjin no saigaikan.* Ibaraki-ken, Koga-shi: Koga reksihi hakubutsukan, 1999.

Köhn, Stephan. "Between Fiction and Non-Fiction: Documentary Literature in the Late Edo Period." In Susanne Formanek and Sepp Linhart, *Written Texts—Visual Texts: Woodblock-Printed Media in Early Modern Japan.* Amsterdam: Hotei Publishing, 2005, 283–310.

Kojima Tōzan and Tōrōan-shujin. *Jishinkō.* Kyoto: Saiseikan, 1830.

Kojima Yoshiyuki. "Chūgoku no daichi no namazu." In Miyata Noboru and Takada Mamoru, eds., *Namazue: Shinsai to Nihon bunka.* Ribun shuppan, 1995, 186–189.

Kojiruien (Dictionary of Historical Terms) Database. Kyoto: Nichibunken, International Research Center for Japanese Studies, 2007. http://www .nichibun.ac.jp/graphicversion/dbase/kojirui_e.html (accessed December 21, 2011).

Konta Yōzō. "Bakumatsu masu-media jōhō." In Miyata Noboru and Takada Mamoru, eds., *Namazue: shinsai to Nihon bunka.* Ribun shuppan, 1995, 77–94.

———. "Edo no saigai jōhō." In Nishiyama Matsunosuke, ed., *Edo chōnin no kenkyū.* Yoshikawa kōbunkan, 1980, 171–282.

Kornicki, Peter. *The Book in Japan: A Cultural History from the Beginning to the Nineteenth Century.* Leiden: Brill, 1998.

Koshimura Shun'ichi. "Sanriku chihō no tsunami saigai gaiyō." In Chūō bōsai kaigi, *1896 Meiji Sanriku jishin tsunami hōkokusho,* 2005, 3–6.

Koshimura Shun'ichi and Shutō Nobuo. "Meiji Sanriku jishin tsunami." In Chūō bōsai kaigi, *1896 Meiji Sanriku jishin tsunami hōkokusho,* 2005, 7–30.

Kotô, Bundjiro [Kotō Bunjirō]. "On the Cause of the Great Earthquake in Central Japan, 1891." *Tōkyō teikoku daigaku kiyō, rika* 5, no. 10 (1893), 295–353.

Laudan, Rachel. *From Mineralogy to Geology: The Foundations of a Science, 1650–1830.* Chicago: University of Chicago Press, 1987.

Ludwin, R. S., and G. J. Smits. "Folklore and Earthquakes: Native American Oral Traditions from Cascadia Compared with Written Traditions from Japan." In L. Piccardi and W. B. Masse, eds., *Myth and Geology.* London: Geological Society, 2007, 67–91.

Lyell, Charles. *Principles of Geology, Being an Attempt to Explain the Former Changes of the Earth's Surface, by Reference to Causes Now in Operation,* 2nd ed., 3 vols. London: John Murray, 1832.

Markus, Andrew L. "Gesaku Authors and the Ansei Earthquake of 1855." In Dennis Washburn and Alan Tansman, eds., *Studies in Modern Japanese Literature.* Ann Arbor: Center for Japanese Studies, University of Michigan, 1997: 53–87.

"Massive Tsunami Projected: Panel Forecasts Nankai Trough Quakes Could Affect 11 Prefectures." *Daily Yomiuri Online,* April 2, 2012.

Mauch, Christof. "Introduction." In Christof Mauch and Christian Pfister, eds., *Natural Disasters, Cultural Responses: Case Studies toward a Global Environmental History.* New York: Lexington Books, 2009.

Mauch, Christof, and Christian Pfister, eds. *Natural Disasters, Cultural Responses: Case Studies toward a Global Environmental History.* New York: Lexington Books, 2009.

McClain, James L. "Edobashi: Power, Space, and Popular Culture in Edo." In James L. McClain, John W. Merriman, and Ugawa Kaoru, eds., *Edo and Paris: Urban Life and the State in the Early Modern Era.* Ithaca, NY: Cornell University Press, 1994, 106–131.

McClain, James L., John W. Merriman, and Ugawa Kaoru, eds. *Edo and Paris: Urban Life and the State in the Early Modern Era.* Ithaca, NY: Cornell University Press, 1994.

Metzler, Mark, and Gregory Smits. "Introduction: The Autonomy of Market Activity and the Emergence of *Keizai* Thought." In Bettina Gramlich-Oka and Gregory Smits, eds., *Economic Thought in Early Modern Japan.* Leiden: Brill, 2010, 1–19.

Miki Haruo. *Kyōto daijishin.* Shibunkaku shuppan, 1979.

Minami Kazuo. *Edo no fūshiga.* Yoshikawa kōbunkan, 1997.

Miura Baien. *Zeigo.* In Yamada Keiji, ed., trans., *Miura Baien.* Nihon no meicho 20, Chūō kōronsha, 1984, 413–610.

Miyata Noboru. *Kinsei no hayarigami.* Hyōronsha, 1972.

———. "Toshi minzokugaku kara mita namazu shinkō." In Miyata Noboru and Takada Mamoru, eds., *Namazue: Shinsai to Nihon bunka.* Ribun shuppan, 1995, 24–33.

———. "'Yonaoshi' to Miroku shinkō: Nihon ni okeru 'yonaoshi' no minzokuteki imi." *Minzokugaku kenkyū* 33, no. 1 (1968): 32–44.

Miyata Noboru and Takada Mamoru, eds. *Namazue: Shinsai to Nihon bunka.* Ribun shuppan, 1995.

Mizuno Shōji. "Chūsei no saigaikan." In Kitahara Itoko, ed., *Nihon saigaishi.* Yoshikawa kōbunkan, 2006, 97–109.

Mukherjee, Saumitra, ed. *Earthquake Prediction.* Leiden: Brill, 2006.

Murayama Masataka. *Shinden kōsetsu* [1856]. In Edo josei bunko, vol. 49. Ōzorasha, 1994 (no pagination).

Musha Kinkichi. *Jishin namazu.* Meiseki shoten, 1995 (originally 1957).

Nagura Tetsuzō. *Etoki, bakumatsu fūshiga to tennō.* Kashiwa shobō, 2007.

———. *Fūshigan ishin henkaku: Minshū wa tennō o dō miteita ka.* Kōsō shobō, 2004.

Nakamura Misao. "Ansei Edo jishin." In Chūō bōsai kaigi, *1855 Ansei Edo jishin hōkokusho.* Fuji sōgō kenkyūsho, 2004, 1–39.

Nakamura Misao, Kayano Ichirō, and Matsuura Ritsuko. "Ansei Edo jishin no shubuken de no higai." *Rekishi jishin* 19 (2003): 32–37.

Nakamura Misao, Matsuura Ritsuko, Kayano Ichirō, Karakama Ikuo, and Nishiyama Akihito. "Ansei Edo Jishin (1855/11/11) no Edo shichū no higai." *Rekishi jishin* 18 (2002): 77–96.

Nakano Misao. *Nishikie-e igaku minzokushi.* Kaneharu shuppan, 1980.

Nakayama Einosuke. "Ansei no kyodai jishin to kawaraban." In Inagaki Funio, ed., *Edo no taihen: Jishin, kaminari, kaji, kaibutsu.* Heibonsha, 1995, 69–77.

"Namazu ga ugoku to kanarazu jishin ga okoru." *Yomiuri shinbun,* April 1, 1932, morning edition, 7.

"Namazu no ugoki de jishin o yochi." *Yomiuri shinbun,* October 14, 1932, morning edition, 4.

Nenzi, Laura. *Excursions in Identity: Travel and the Intersection of Place, Gender, and Status in Edo Japan.* Honolulu: University of Hawai'i Press, 2008.

Nienhausser, William H. Jr., ed. *The Grand Scribe's Records,* vol. 1: *The Basic Annals of Pre-Han China by Ssu-ma Ch'ien.* Trans. Tsai-fa Cheng et al. Bloomington: Indiana University Press, 1994.

Nihon gakushi-in, eds. *Meiji-zen Nihon butsuri kagakushi.* Nihon gakjutsu shinkōkai, 1964.

Nishikawa Joken. *Kaii bendan,* 8 vols. Kyoto: Ryūshiken, 1715.

Nishimaki Kōzaburō, ed. *Kawaraban shinbun: Edo, Meiji sanbyaku jiken,* vol. 1. Heibonsha, 1978.

Nishiyama Akihito. "Ansei Nankai jishin ni okeru Ōsaka de no shinsai taiō." In Chūō bōsai kaigi, *1854 Ansei Tōkai jishin, Ansei Nankai jishin hōkokusho,* 2005, 42–67.

———. "Kyōto de no higai to jishin taiō." In Chūō bōsai kaigi, *1662 Kanbun Ōmi-Wakasa jishin hōkokusho.* Mizuho Jōhō sōken kabushikigaisha, 2005, 124–157.

Noguchi Takehiko. *Ansei Edo jishin: Saigai to seiji kenryoku.* Chikuma shobō, 1997.

Okada Yoshimitsu. *Saishin Nihon no jishin chizu.* Tōkyō shoseki, 2006.

Okumiya Masaaki. *Kokuryōki* (no date or publisher).

Oliver-Smith, Anthony. "Theorizing Disasters: Nature, Power, and Culture." In Susanna Hoffman and Anthony Oliver-Smith, eds., *Catastrophe and Culture: The Anthropology of Disaster.* Santa Fe, NM: School of American Research Press, 2002, 18–34.

Oliver-Smith, Anthony, and Susanna M. Hoffman, eds. *The Angry Earth: Disaster in Anthropological Perspective.* New York: Routledge, 1999.

Ooms, Herman. *Imperial Politics and Symbolics in Ancient Japan: The Tennmu Dynasty, 650–800.* Honolulu: University of Hawai'i Press, 2009.

———. *Tokugawa Ideology: Early Constructs, 1570–1680.* Princeton, NJ: Princeton University Press, 1985.

———. *Tokugawa Village Practice: Class, Status, Power, Law.* Berkeley: University of California Press, 1996.

Ōsaki shoji, ed. *Kinsei Nihon tenmon shiryō.* Hara shobō, 1994.

Ouwehand, Cornellis. *Namazu-e and Their Themes: An Interpretative Approach to Some Aspects of Japanese Folk Religion.* Leiden: E. J. Brill, 1964.

Piccardi, L., and W. B. Masse, eds. *Myth and Geology.* London: Geological Society, 2007.

"Q: Sen-nen ni ichido no kyodaijishin to iwareru Higashi Nihon dai-shinsai, naze sen-nen ni ichido na no? A: Jōgan jishin no sairai toiu mikata." *Yomiuri Online,* October 20, 2011.

Quenet, Gregory. "Earthquakes in Early Modern France: From the Old Regime to the Birth of New Risk." In Andrea Janku, Gerrit Schenk, and Franz Mauelshagen, eds., *Historical Disasters in Context: Science, Religion, and Politics.* New York: Routledge, 2012, 94–114.

Ravina, Mark J. "Confucian Banking: The Community Granary (*Shasō*) in Rhetoric and Practice." In Bettina Gramlich-Oka and Gregory Smits, eds., *Economic Thought in Early Modern Japan.* Leiden: Brill, 2010, 179–204.

———. *Land and Lordship in Early Modern Japan.* Stanford, CA: Stanford University Press, 1999.

Reilly, Benjamin. *Disaster and Human History: Case Studies in Nature, Society, and Catastrophe.* Jefferson, NC: McFarland & Company, 2009.

Rikitake Tsuneji, "Deeta ni miru namazu to jishin." In Miyata Noboru and Takada Mamoru, eds., *Namazu-e: Shinsai to Nihon bunka.* Ribun shuppan, 1995, 148–156.

Roberts, Luke S. *Performing the Great Peace: Political Space and Open Secrets in Tokugawa Japan.* Honolulu: University of Hawai'i Press, 2012.

Rubinger, Richard. *Popular Literacy in Early Modern Japan.* Honolulu: University of Hawai'i Press, 2007.

Saitō Gesshin. *Ansei itsubō bukō chidō no ki.* In Edo sōsho kankō kai, eds., *Edo sōsho 9.* Edo sōsho kankōkai, 1917, 2–66 (pagination separate for each work in the volume).

———. *Bukō nenpyō,* 2 vols. Tōyō bunko #116, 118. Compiled and edited by Kaneko Mitsuharu. Heibonsha, 1968, 1992.

Saitō Masachi. *Goze kudoki jishin no minoue.* Publisher unknown, 1829.

Sangawa Akira. *Jishin no Nihonshi: Daichi wa nani o kataru no ka?* Chūōkōron shinsha, 2007.

Sasaki Junnosuke. *Yonaoshi.* Iwanami shoten, 1979.

Satō Hiroo. *Shinkoku Nihon* (Chikuma shinsho 591). Chikuma shobō, 2006.

———. "Wrathful Deities and Saving Deities." In Mark Teeuwen and Fabio Rambelli, eds., *Buddhas and Kami in Japan: Honji suijaku as a Combinatory Paradigm.* New York: RoutledgeCurzon, 2003, 95–114.

Sayama Mamoru. *Ansei Edo jishin saigaishi,* 2 vols. Kaiji shoin, 2004. (Reprint of a 1968 edition published by Tōkyō-to, Sōmukyoku, Gyōseibu.)

Schencking, J. Charles. "The Great Kantō Earthquake and the Culture of Catastrophe and Reconstruction in 1920s Japan." *Journal of Japanese Studies* 34, no. 2 (summer 2008): 295–331.

Seigle, Cecilia Segawa. *Yoshiwara: The Glittering World of the Japanese Courtesan.* Honolulu: University of Hawai'i Press, 1993.

Shearer, Peter M. *Introduction to Seismology,* 2nd ed. New York: Cambridge University Press, 2009.

Shiga, Hidemi. "The Catfish Underground: Japan's Earthquake Folklore and Popular Responses to Disaster." *Orientations Magazine for Collectors and Connoisseurs of Asian Art* (April 2006): 77–80.

Shimamura Hideki. *"Jishin yochi" wa uso darake.* Kōdansha, 2008.

———. *Nihonjin ga shiritai jishin no gimon 66: Jishin ga ōi Nihon dakarakoso chishiki no sonae mo wasurezuni!* Sofutobanku kurieitibu, 2008.

Shimizu Isao. *Edo no manga.* Kōdansha, 2003.

———. *Manga no rekishi.* Iwanami shoten, 1991.

Shinano mainichi shinbunsha kaihstsukyoku shuppanbu, ed. *Kōka yonen Zenkōji daijishin.* Nagano-shi: Shinano mainichi shinbunsha, 1977.

"Shinkaigyo, Shizuoka-ken de aitsugi shutsugen . . . daijishin no maebure?" *Yomiuri shinbun,* June 9, 2012.

Shinsai yobō chōsakai, eds. *Dai-Nihon jishin shiryō,* 2 vols. Shibunkaku, 1973.

"Shōmu tennō, Bukkyō e no fukai kie—kikkake wa daijishin?" *Asahi shinbun,* March 3, 2006.

Smith, Henry D., II. "The History of the Book in Edo and Paris." In James L. McClain, John W. Merriman, and Ugawa Kaoru, eds., *Edo and Paris: Urban Life and the State in the Early Modern Era.* Ithaca, NY: Cornell University Press, 1994, 332–352.

Smits, Gregory. "Conduits of Power: What the Origins of Japan's Earthquake Catfish Reveal about Religious Geography." *Japan Review* 24 (2012): 41–65.

———. "Danger in the Lowground: Historical Context for the March 11, 2011 Tōhoku Earthquake and Tsunami." *Asia-Pacific Journal* 9, issue 20, no. 4 (May 16, 2011). http://www.japanfocus.org/-Gregory-Smits/3531. Reprinted at History News Network (HNN), Wednesday, May 18, 2011. Reprinted in *Engineering World* (June–July 2011): 16–21. Reprinted as "Building on Fault Lines" in Jeff Kingston, ed., *Tsunami, Japan's Post-Fukushima Future: Essays on the Aftershocks of Japan's Nuclear Nightmare* (Foreign Policy, 2011): 67–87.

———. "*Namazu-e*: Catfish Prints of 1855." *Andon* (Publication of the Society for Japanese Arts) 86 (2009): 35–46.

———. "Shaking Up Japan: Edo Society and the 1855 Catfish Picture Prints." *Journal of Social History* 39, no. 4 (summer 2006): 1045–1077.

———. "Warding off Calamity in Japan: A Comparison of the 1855 Catfish Prints and the 1862 Measles Prints." *East Asian Science, Technology, and Medicine* 30 (2009), 9–31.

Steele, M. William. *Alternative Narratives in Japanese History.* New York: RoutledgeCurzon, 2003.

Stein, Seth. *Disaster Deferred: How New Science Is Changing Our View of Earthquake Hazards in the Midwest.* New York: Columbia University Press, 2010.

Suehiro Sachiyo. "Ōtsue no hyōtan-namazu." In Miyata Noboru and Takada
    Mamoru, eds., *Namazue: Shinsai to Nihon bunka.* Ribun shuppan,
    1995, 182–185.
Sugimoto Isao, ed. *Kagakushi.* Taikei Nihonshi sōsho 19, Yamakawa
    shuppan, 1976.
Suyehiro [Suehiro], Yasuo. "Some Observations on the Unusual Behavior of Fishes
    Prior to an Earthquake." Earthquake Research Institute, Tokyo Imperial
    University, March 1934, 228–231.
Suzuki Tōzō and Koike Shōtarō, eds. *Fujiokaya nikki.* Kinsei shomin seikatsu
    shiryō, vol. 15. San'ichi shobō, 1995.
Tahara Tsugio, ed. *Yamaga Sokō.* Nihon no meicho 12. Chūō kōronsha, 1983.
Tahara Tsugio and Morimoto Jun'ichirō, eds. *Yamaga Sokō.* Nihon shisō taikei 32.
    Iwanami shoten, 1970.
Taikyoku jishinki. In Aoki Kunio et al., eds., *Taikyoku jishinki, Ansei kenbunroku,
    Jishin yobōsetsu, Bōkasaku zusetsu.* Edo kagaku koten sōsho, vol. 19. Kōwa
    shuppan, 1979.
Takada Mamoru. "Namazue no chosakutachi: Chosha, gakō o meguru
    bakumastu bunka jōkyō." In Miyata Noboru and Takada Mamoru, eds.,
    Namazue: Shinsai to Nihon bunka. Ribun shuppan, 1995, 34–51.
Takagi Shunsuke. *Ee ja nai ka.* Kyōikusha, 1979.
Takeuchi, Makoto. "Festivals and Fights: The Law and the People of Edo." In James
    L. McClain, John W. Merriman, and Ugawa Kaoru, eds., *Edo and Paris: Urban
    Life and the State in the Early Modern Era.* Ithaca, NY: Cornell University
    Press, 1994, 284–406.
Tenkei wakumon, vol. 2, "Jishin." Osaka: Ōsaka shobō (1794 revision of the 1730
    ed., 3 vols.).
Terada, Torahiko. "Luminous Phenomena Accompanying Destructive Sea-Waves
    (Tunami)." Earthquake Research Institute, Tokyo Imperial University, March
    1934, 25–35. http://hdl.handle.net/2261/13770 (accessed January 2, 2011).
Thomas, Julia Adeney. *Reconfiguring Modernity: Concepts of Nature in Japanese
    Political Ideology.* Berkeley: University of California Press, 2001.
Thompson, Sarah E., and H. D. Harootunian. "The Politics of Japanese Prints." In
    Sarah E. Thompson and H. D. Harootunian, *Undercurrents of the Floating
    World: Censorship and Japanese Prints.* New York: Asia Society Galleries,
    1991: 29–91.
———. *Undercurrents of the Floating World: Censorship and Japanese Prints.* New
    York: Asia Society Galleries, 1991.
Tōkyō daigaku jishin kenkyūjo, ed. *Shinshū Nihon jishin shiryō,* vol. 5,
    supplement 2 (*Ansei Edo jishin*), parts 1 and 2. Nihon denki kyōkai, 1985.
———. *Shinshū Nihon jishin shiryō,* vol. 5, supplement 6 (*Zenkōji jishin*), parts 1
    and 2. Tōkyō daigaku jishin kenkyūjo, 1988.

————. *Shinshū Nihon jishin shiryō, hoi* (supplement). Nihon denki kyōkai, 1989.

Tōkyō kagaku hakubutsukan, ed. *Edo jidai no kagaku.* Meicho kankōkai, 1980 (originally published 1934).

Tōkyō shiyakusho, ed. *Tōkyōshi shikō, kyūsaihen 4.* Rinsen shoten, 1975 (originally published 1922).

Tomisawa Tatsuzō. "'Namazu-e no sekai' to minzoku ishiki." *Nihon minzokugaku* 207 (1996): 85–105.

————. "Nishiki-e no nyūsu sei: Namazue, hashikae, Bōshin sensō-ki no fūshiga o megutte." In Kinoshita Naoyuki and Yoshimi Shunya, eds., *Nyūsu no tanjō: Kawaraban to shinbun nishikie no jōhō sekai.* Tōkyō daigaku sōgō kenkyū hakubutsukan, 1999, 192–203.

Toyo Tokinari. *Honchō jishinki.* In Edo josei bunko, vol. 49. Ōzorasha, 1994 (no pagination).

Tributsch, Helmut. *When the Snakes Awake: Animals and Earthquake Prediction.* Trans. Paul Langner. Cambridge, MA: MIT Press, 1982.

Tsu, Timothy Yun Hui. "Making Virtues of Disaster: 'Beautiful Tales' from the Kobe Flood of 1938." *Asian Studies Review* 32, no. 3 (June 2008): 197–214.

Tsuchiura shiritsu hakubutsukan, ed. *Namazue kenbunroku: Ō-Edo bakumatsu namazue jijō* (Tsuchiura Shiritsu Hakubutsukan dai 17-kai tokubetsuten). Tsuchiura, Japan: Tsuchiura shiritsu hakubutsukan, 1996.

Tsuji Yoshinobu. "Ansei Tōkai, Nankai jishin no jitsuzō to senjin no saigai kyōkun." In Chūō bōsai kaigi, *1854 Ansei Tōkai jishin, Ansei Nankai jishin hōkokusho.* Chūō bōsai kaigi, 2005, 1–6.

————. *Sennen shinsai: Kurikaesu jishin to tsunami no rekishi ni manabu.* Daiyamondo sha, 2011.

————. *Zukai, Naze okoru? Itsu okoru? Jishin no mekanizumu.* Nagaoka shoten, 2010.

Tsukahara Hiroaki, "Zenkōji jishin wa dono yōnishite hasseishita ka." In Akahane Sadayuki and Kitahara Itoko, eds., *Zenkōji jishin ni manabu.* Nagano-shi: Shinano mainichi shinbunsha, 2003, 35–47.

Tsukuda Tameshige. *Jishin yochi no saishin kagaku: Hasshō no mekanizumu to yochi kenkyū no saizensen.* Sofutobanku kurieitibu, 2007.

Ueda Kazue and Usami Tatsuo. "Yūshi irai no jishin kaisū no hensen." *Rekishi jishin* 6 (1990): 181–187.

Ulusoy, Ülkü, and Himansu Kumar Kundu, eds. *Future Systems for Earthquake Early Warning.* New York: Nova Science Publishers, 2008.

Unno Kazutaka. "Cartography in Japan." In J. B. Harley and David Woodward, eds., *The History of Cartography,* vol. 2, book 2: *Cartography in the Traditional East and Southeast Asian Societies.* Chicago: University of Chicago Press, 1994.

————. "Kigan, majinai no tsukawareta Nihonzu." In Unno Kazutaka, *Tōyō chirigakushi kenkyū, Nihon hen.* Osaka: Seibundō shuppan, 2005, 215–255.

"Maps of Japan Used in Prayer Rites or as Charms" is a subset of this longer article.

——. "Maps of Japan Used in Prayer Rites or as Charms." *Imago Mundi* 46 (1994), 65–83.

Usami Tatsuo. "Kaisetsu." In Aoki Kunio et al., eds., *Taikyoku jishinki, Ansei kenmonroku, Jishin yobōsetsu, Bōkasaku zusetsu.* Edo kagaku koten sōsho, vol. 19. Kōwa shuppan, 1979, 31–47.

——. *Nihon higai jishin sōran [416]–2001, saishin-ban* (English title: Materials for Comprehensive List of Destructive Earthquakes in Japan [416]–2001, [latest ed.]). Tōkyō daigaku shuppankai, 2003.

——, ed. *"Nihon no rekishi jishin shiryō" shūi: Saikō 2-nen yori Shōwa 21-nen ni itaru,* vol. 3. Watanabe tansa gijutsu kenkyūjo, 2005.

——, ed. *"Nihon no rekishi jishin shiryō" shūi: Seimu Tennō 3-nen yori Shōwa 39-nen ni itaru,* vol. 2. Yamato tansa gijutsu kabushikigaisha, 2002.

——, ed. *"Nihon no rekishi jishin shiryō" shūi: Tenmu Tennō rokunen yori Ansei gannen ni itaru,* vol 4. Watanabe tansa gijutsu kenkyūjo, 2008.

——, ed. *"Nihon no rekishi jishin shiryō" shūi: Tenmu Tennō 13-nen yori Shōwa 58-nen ni itaru.* Nihon denki kyōkai, 1998.

Vaporis, Constantine Nomikos. *Tour of Duty: Samurai, Military Service in Edo, and the Culture of Early Modern Japan.* Honolulu: University of Hawai'i Press, 2008.

Vlastos, Stephen. *Peasant Protests and Uprisings in Tokugawa Japan.* Berkeley: University of California Press, 1986.

Wakabayashi, Bob Tadashi. *Anti-Foreignism and Western Learning in Early-Modern Japan: The New Theses of 1825.* Cambridge, MA: Council on East Asian Studies, Harvard University, 1986.

Wakamizu Suguru. *Edokko kishitsu to namazue.* Kadokawa gakugei shuppan. 2007.

——. *Namazu wa odoru: Edo no namazue omoshiro bunseki.* Bungeisha, 2003.

Walthall, Anne. "Edo Riots." In James L. McClain, John W. Merriman, and Ugawa Kaoru, eds., *Edo and Paris: Urban Life and the State in the Early Modern Era.* Ithaca, NY: Cornell University Press, 1994, 407–428.

——, ed., trans. *Peasant Uprisings in Japan.* Chicago: University of Chicago Press, 1991.

——. *Social Protest and Popular Culture in Eighteenth-Century Japan.* Tucson: Association of Asian Studies, University of Arizona Press, 1986.

Weiner, Michael A. *The Origins of the Korean Community in Japan, 1910–1923.* Manchester, UK: Manchester University Press, 1989.

Wigen, Kären. *A Malleable Map: Geographies of Restoration in Central Japan, 1600–1912.* Berkeley: University of California Press, 2010.

Wilson, George. *Patriots and Redeemers in Japan: Motives in the Meiji Restoration.* Chicago: University of Chicago Press, 1982.

Yamada Keiji, ed., trans. *Miura Baien.* Nihon no meicho 20, Chūō kōronsha, 1984.

Yamamoto, Daiki. "Sea Serpents' Arrival Puzzling, or Portentous." *Japan Times,* March 6, 2010.

Yamamoto Takeo. "Shiryō ginmi no hitsuyōsei." In Hagiwara Takahiro et al., *Kojishin: Rekishi shiryō to katsu-dansō kara saguru.* Tōkyō daigaku shuppankai, 1982, 39–59.

Yamanaka, Chihiro, Hiroshi Asahara, Yutaka Emoto, and Yuko Esaki. "Earthquake Precursors: From Legends to Science and a Possible Early Warning System." In Ülkü Ulusoy and Himansu Kumar Kundu, eds., *Future Systems for Earthquake Early Warning.* New York: Nova Science Publishers, 2008, 201–208.

Yamashita Fumio. *Tsunami no kyōfu: Sanriku tsunami denshōroku.* Sendai, Japan: Tōhoku daigaku shuppankai, 2005.

———. *Tsunami tendenko: Kindai Nihon no tsunamishi.* Shin Nihon shuppansha, 2008.

Yamazaki Bisei, comp. *Ōjishin rekinenn kō* [1856]. In Edo josei bunko, vol. 49. Ōzorasha, 1994 (no pagination).

Yampolsky, Philip B., ed. Burton Watson et al., trans. *Selected Writings of Nichiren.* New York: Columbia University Press, 1990.

Yoshimura Akira. *Sanriku kaigan ōtsunami.* Bungei shunjū, 2004.

Zeilinga de Boer, Jelle, and Donald Theodore Sanders. *Earthquakes in Human History: The Far-Reaching Effects of Seismic Disruptions.* Princeton, NJ: Princeton University Press, 2005.

INDEX

Abe Masahiro, 30, 126–127, 142–143, 148, 167
Account of Broken Windows (Mado yabure no ki), 46, 77, 113
Account of Chastisement and Shaking (Chōshin hiroku), 60–61, 83, 95
Accounts of the Ansei Earthquake (Ansei itsubo jishin kibun), 114
aftershocks, 3, 14, 15–16, 31–32, 50, 52–52, 55–56, 57, 64, 66, 76–77, 82, 104–105, 140, 178
After the Shaking (Nai no nochimigusa), 47, 104, 116, 118
Aizawa Seishisai, 156–157, 159
Alpha and Omega Explained (Kenkon bensetsu), 43–44
Amaterasu, 3, 20, 74, 99, 142–144, 146, 148, 154–161, 168–170
Anaximenes of Miletus, 43
animal behavior and earthquakes, 1–2, 60, 187–188. *See* catfish: and earthquakes; fish
Ansei Chronicle (Ansei kenmonshi), 32, 53, 56–57, 60, 62–63, 82, 84–86, 114, 132–133, 149, 166, 187, 189–190
Ansei Iga earthquake. *See* Iga-Ueno earthquake
Ansei Record (Ansei kenmonroku), 47, 53, 65, 78–79, 82, 104, 158, 187, 189–190
Ansei Tōkai and Nankai earthquakes (1854), 10, 15, 31, 96–97, 101, 139, 150, 151, 165, 175, 192
architecture, Meiji-era debates about, 107

Aristotle, 43–45
artillery batteries (*daiba*), 74–75, 110, 115–117, 137, 141, 146–147, 159, 166, 168
Asai Ryōi. *See Foundation Stone*
Astronomy Questions Answered (Japanese: *Tenkei wakumon*; Chinese: *Tianjing huowen*), 44–46, 49, 51, 53, 65
Atago Gongen, 146, 157, 169
attendance on shogun (*sankin-kōtai*) requirements relaxed, 127, 137, 146–147, 168, 170

bakufu: attempts to control prices and wages, 24, 120, 124–125, 131; nature of its power, 23–24, 84, 145–147; relations with retainers after Ansei Edo earthquake, 126–128, 132, 146–147; relief to townspeople after Ansei Edo earthquake, 128–132. *See also* disaster relief
bakuhan state, 3, 20, 147, 157, 170. *See also* Japan; conceptions of
Benevolent King Sutra (Ninnō-kyō), 41
Biwakō seigan earthquake. *See* Kanbun earthquake
Borland, Janet, 189–190
Brahe, Tycho, 44
buddhas. *See kami*: and buddhas
Buddhist cosmology, 32, 38, 40–44
Bukō Chronicle (Bukō nenpyō), 64

casualty figures, Ansei Edo earthquake, 104, 113–114, 158, 177, 180, 181